Tensor Calculus with Object-Oriented Matrices for Numerical Methods in Mechanics and Engineering

Udo F. Meissner

Tensor Calculus with Object-Oriented Matrices for Numerical Methods in Mechanics and Engineering

Fundamentals and Functions for Tensor/Matrix Algorithms of the Finite Element Method

Udo F. Meissner
Erkrath, Nordrhein-Westfalen, Germany

ISBN 978-3-031-59301-7 ISBN 978-3-031-59302-4 (eBook)
https://doi.org/10.1007/978-3-031-59302-4

This Springer imprint is published by the registered company Springer Nature Switzerland AG
The registered company address is: Gewerbestrasse 11, 6330 Cham, Switzerland

To my honored academic mentor

Prof. Dr.-Ing. Dieter Withum
** September 29, 1928   † June 25, 1979*

Full Professor of Fluid Mechanics
and Electronic Computing in Civil Engineering
at the Technical University of Hannover from 1967 to 1974,
Full Professor of Mechanics and Statics
at the University of the Bundeswehr, Munich from 1974 to 1979

dedicated to his memory.

Udo F. Meissner

Preface to the Updated Edition

In the first edition, the concepts of matrix and tensor calculus are structured and explained according to object-oriented principles [Meissner 2022]. By implementing these concepts in C++ algorithms, the matrix classes were realized, and their associated methods were illustrated through typical numerical examples.

To further enhance the overview of methods used in various equation solvers, the current edition now introduces the qMATRIX matrix class. This addition streamlines the handling of linear equation systems by providing a procedural approach to matrix division.

Given the challenge of visualizing the exact positions of tensor objects and the arrangement of reference systems in three-dimensional Euclidean space, the clarity of complex relationships is now enhanced through graphical methods for spatial object presentation.

Consequently, an additional chapter on linear algebra is included, featuring the perspective representation of treated objects translated into graphical illustrations. Through parallel and central perspectives, various objects such as points, lines, and vectors can be depicted in three dimensions, facilitating a better understanding of their characteristics.

This approach is particularly effective with stereo perspective, where objects emerge from two-dimensional image projections, appear to float in space, and can be viewed from different angles. To implement this technique, complementary tensor and graphic classes are provided and utilized for perspective viewing of basis and vector objects in Euclidean space.

The presentations in the book offer a comprehensive overview of mathematical fundamentals and characteristic numerical methods, particularly in the context of finite elements, drawing from my research and teaching experiences at the Universities of Hannover and Darmstadt [1,2,3]. I extend special thanks to our staff members *Dr.-Ing. Siegfried Gaertner* [1] and *Dr.-Ing. Hermann Meyer* [2], as well as *Dr.-Ing. Jochen Ruben* [3], *Dipl.-Ing. Inke Terlinden* [3], and *Dr.-Ing. Michael Burghardt* [3], for their exemplary dedication and successful delivery of matrix/tensor calculus contents to advanced students in the Structural Engineering and Numerical Methods teaching programs.

Preface to the 1st Edition

The inspiration for this book dates back to the 1960s and 1970s when new algorithms and program systems for finite element models in structural engineering were developed at the Technical University of Hannover in the field of Electronic Computing in Civil Engineering. These developments, based on additional support from the German Research

[1] University Hannover, Institute of Fluid Mechanics and Electronic Computing in Civil Engineering
[2] University Hannover, Institute of Mechanics and Computational Mechanics
[3] Technical University Darmstadt, Institute of Numerical Methods and Informatics in Civil and Structural Engineering

Foundation (DFG) for Special Research Areas, extended to broad research in coastal engineering areas. Building upon the theoretical foundations of mechanics/statics and fluid mechanics, classical tensor and matrix calculus methods were utilized to implement finite element algorithms using powerful simulation tools such as the MECCA program system [Herrling/Pfeiffer/Beyer 1979] on institute computers and mainframes at the Regional Computing Center of Lower Saxony. Comprehensive program libraries for numerical computation with matrices and for managing digital data were created, sometimes initially coded in assembler for reasons of run time optimization, later transitioning to ALGOL and FORTRAN programming languages [Meissner 1978]. These efforts enabled the development of digital models and realistic simulations for large physical application domains [Ehlers 1971], [Meissner 1971], [Grotkop 1972], [Schulze 1973], with contributions from [Beyer 1977], [Ebeling 1977], and [Herrling 1977] to the development of MECCA.

Even back then, Withum pursued an innovative approach, building on the work of [Duschek/Hochrainer 1968], to develop a unified methodology for processing tensor quantities and matrices by storing them in multidimensional arrays of programming languages. Lectures and seminars were conducted, leading to concepts and manuscripts for implementation in teaching and research. However, procedural programming languages at the time did not allow for specialized syntax and semantics of matrix operations. As a result, the approach relied on custom memory management and specific program routines that were highly efficient, which were eventually ported to MIMD computers successfully in the 1990s by the Institute for Numerical Methods and Computer Science in Civil and Structural Engineering at the Technical University of Darmstadt with support of the German Research Foundation (DFG) [Laemmer 1996], [Meissner 2006].

It wasn't until the late 1980s that software development concepts improved with the introduction of the object-oriented paradigm in process modeling and software engineering. This paradigm enabled matrices to be treated as objects, structuring their specific properties within classes. With the object-oriented paradigm, matrix operations could be bound to the objects themselves, allowing the usual arithmetic and functional operators to be overloaded with matrix calculus operators. This approach was implemented at the Technical University of Darmstadt in the 1990s, introduced and successfully tested in teaching of the Institute for Numerical Methods and Computer Science in Civil and Structural Engineering. This led to a coherent concept for utilizing tensor calculus in numerical methods and adapting syntax and semantics for implementation in object-oriented software modules based on the C++ programming language. Tensor objects could now be directly used in numerical algorithms, especially for the application of the Finite Element Method (FEM) in mechanics and engineering.

With the publication of this book, a guide is presented for this comprehensive concept, based on a consistent notation for processing tensors, matrices, matrix objects, and FEM algorithms. Well-known matrix operations can thus be intuitively used and applied as tools of matrix arithmetic in teaching, research, and professional practice.

I would like to express my sincere thanks to *Dipl.-Ing. Eberhard Pfeiffer*, Academic Director emeritus at the University of the Bundeswehr Munich, former assistant at the chairs of Prof. Withum, and custodian of the Dieter Withum Archive, for his advice in developing the technical content. My special thanks also go to *Barbara Hartmann*, Rector emeritus, for the editorial work on the book manuscript. Through this collaborative effort and especially with the supportive assistance of my wife, *Heike Fredrich-Meissner*, this work was successfully completed and brought to publication.

Technical University of Darmstadt [IIB], February 2024 *Udo F. Meissner*

Contents

Chapter 1
Introduction

Abstract

The main goal of the book is to establish a synthesis between classical matrix and tensor methods, and modern software technology. To achieve this, the approach involves the development of a cohesive methodological framework using object-oriented methods. This framework enables the seamless transformation of the theoretical modeling principles of mechanics into numerical computational programs for solving engineering problems. This is done without encountering methodological inconsistencies in various subareas. One classic example of such a challenge is the symbolic notation of tensor and matrix calculus, which features complex syntax and semantics. This notation, originally designed for 2D media, is still widely used today, especially in the form of matrix arithmetic. Despite the availability of multidimensional arrays in programming languages for matrix implementation, this notation remains prevalent.
Consequently, the well-established index notation for tensors and matrices becomes invaluable when specifying array objects within matrix operations. It may appear less popular due to the extensive use of sub- and super-indices. However, when applied with Einstein's summation convention, it becomes highly effective and allows for transparent handling of multidimensional matrix operations, akin to ordinary scalar objects. Therefore, the introductory chapters explicitly introduce both symbolic and index notations to get readers acquainted with both notations in detail.
Index notation has grown immensely important for implementing tensor and matrix methods in object-oriented software designs. Multidimensional objects can be effectively implemented using specialized class structures in programming languages like C++ or Java. Compared to algorithmic programming languages such as C or FORTRAN, the object-oriented paradigm offers several advantages in declaring and defining tensor and matrix objects, as detailed by [Breymann 1997].
Some of the key topics covered in the book include:
- Declaration and definition of new class members in structured and related classes,
- Encapsulation of class member data and functions in private or protected areas (hidden and not accessible to the user),
- Definition of public interfaces for communication with class members (e.g., by calling functions),
- Inheritance of properties between class relations,
- Dynamic generation and deletion of new class instances during run time,
and additionally, C++ allows for:
- Overloading of arithmetic and functional operators as per its language specifications.

Using these features, matrix operations can be implemented similarly to familiar scalar operations. These aspects of the approach are described and utilized in Chapter 3, considering the relevant properties of matrices and tensors outlined in Chapters 2 and 4. Finally, the practical

U. F. Meissner, *Tensor Calculus with Object-Oriented Matrices for Numerical Methods in Mechanics and Engineering*, https://doi.org/10.1007/978-3-031-59302-4_1

application of the synthesized tensor/matrix methods is demonstrated through two typical software examples for Finite Element Method (FEM) applications in Chapter 7.

The primary objective of the book adheres to the rigorous approach that tensors, widely used in the context of consistent theories of mechanics and physics, should also be effectively utilized for numerical computations through object-oriented matrix methods in engineering. As such, it assumes a strong foundation in matrix and software methods, which are discussed in subsequent chapters.

The intention of the book is to synthesize classical matrix and tensor methods on one hand, and modern software technology and object-oriented methods on the other, for numerical algorithms used in analysis of engineering structures. The goal is to develop a comprehensive methodological concept that seamlessly translates the theoretical foundations of mechanics/statics into numerical calculation programs, without encountering methodological gaps in different areas. A classic example of this is the symbolic notation of tensor and matrix calculus. While this notation is traditionally well-developed, compactly formulated, and descriptively represented, beginners generally struggle to understand how this methodology can be efficiently translated into program code on digital computing systems. The key to overcome this difficulty lies in the object-oriented paradigm, which is widely used in software development, for example, in programming languages like C++ and Java, and its advantages have already been extensively explored in engineering [Hartmann (Hrsg.) 2000]. In Building Information Modeling (BIM), this object-oriented methodology has also proven to be a solid foundation for digital modeling of planned or constructed infrastructure, along with its semantic and logistical information [Rueppel (Hrsg.) 2007] resp. [Borrmann/König/Koch/Beetz (Hrsg.) 2018].

The need for a consistent methodological concept is especially prominent in higher education, where students in various disciplines struggle to integrate diverse fundamentals and methods. Therefore, the present concept aims to bridge these disparities and integrate approaches from tensor algebra, matrix calculus, object-oriented software, and finite element numerics. The author has tested this approach in specialized courses on Numerical Methods and Informatics in Civil and Structural Engineering at the Technical University of Darmstadt, achieving significant success in conveying the subject matter.

A great demand for modern software concepts also exists in professional engineering practice, as numerical methods and computer science are crucial areas for developing and applying practical modeling and calculation systems e.g. for real structural and hydromechanical applications. For instance, since the 1960s, the Finite Element Method (FEM) has reached its current pinnacle in statics and mechanics [Zienkiewicz/Taylor 1989], as well as in dynamics [Clough/Penzien 1975], [Bathe 1986], accompanied by corresponding software systems. This has resulted in powerful software tools, widely available and used across various industrial fields and serving as the foundation for design and construction. The final aim of this book is to address some typical applications from the field of FEM.

The individual chapters serve to comprehensively present the scientific foundations for the intended synthesis of matrix and tensor calculus with the numerical algorithms of object-oriented matrices:

- In the introductory Chapter 2, the most essential fundamentals of classical matrix calculus are compiled, as in [Zurmuehl/Falk 1986] and [Pestel/Leckie 1963]. The primary goal is to introduce the reader to the index notation for multidimensional matrices and explain arithmetic operations. This notation is important to become familiar with, as it significantly differs from the symbolic notation used in theoretical papers. Another crucial aspect is the generalization of the conventional representation from one- and two-dimensional to multidimensional matrices, as this is fundamental for all subsequent considerations. Extending traditional calculation schemes, which originally were designed for use of two-dimensional media, has long been supported to handle multidimensional arrays by common programming languages used in the mathematical and natural sciences. This introduction to index notation also prepares the reader for dealing with the adequate covariant and contravariant tensors in Chapters 4 and 6.
- In Chapter 3, the focus is initially on the structure and storage of multidimensional matrices, which are mapped onto the one-dimensional physical main memory of the conventional digital von Neumann computers. Section 3.2 illustrates these relationships and the possible matrix representations, along with their mathematical mapping rules. Thereafter follows the exemplary integration of vectors and multidimensional matrices into object-oriented matrix classes in the C++ programming language. The primary goal is to illustrate the organization of matrix objects within a class structure according to the principles of encapsulation, inheritance, interfaces, and communication through public functions. Readers may supplement their knowledge of object-oriented programming by studying relevant literature e.g. [Breymann 1997] on C++ resp. [Rumbaugh/Blaha/Premerlani/Eddy/Lorensen 1991] on object-oriented design, as well as by conducting research on the internet. In addition to creating and managing static or dynamic matrix objects, it is particularly important also to recognize the advantage of overloading the in C++ implemented operators for intuitive handling of the new matrix syntax. These aspects will be demonstrated later in Chapter 7 through two typical program functions, PARALLELOGRAM and TRIANGLE, implemented in C++ routines. Through operator overloading, it becomes possible to transfer the familiar syntax of scalar arithmetic and analogous matrix calculations to the new matrix objects in the programming language, greatly enhancing the transparency of working with matrix computations.
- Tensor calculus, with its consistent formulation of invariants and transformations, has held significant importance in engineering since many years. At this point, early works in the mechanics of structures, such as [Green/Zerna 1954], where the formulation of the theory was renovated, and mathematical compendia on tensor calculus in index notation, such as [Duschek/Hochrainer 1968], may serve for basic studies.
  Chapter 4 compiles the fundamental concepts of tensor algebra in this context, primarily to make the approach of object-oriented matrix calculus understandable and to consistently represent the methodology. Thus, the index notation allows the arithmetic operations of tensor calculus to be seamlessly translated into practical programming tools using the new matrix classes and methods.
- In general, it is challenging to intuitively represent elementary tensor quantities, such as basis vectors, metrics, and vectors, which have geometric and mechanical significance, in three-dimensional Euclidean space. To address this, the preceding chapters establish

the corresponding tensor and matrix classes and make them available through program functions. In Chapter 5, additional methods will be introduced to visually illustrate such quantities. Excellent tools for this purpose are graphical representations in the form of spatial perspectives that convey the correlation of different quantities in space and enhance understanding of their relationships. At this point, basic mathematical graphical methods are used to easily achieve projections of three-dimensional objects onto two-dimensional image windows. These methods are presented based on tensor and graphic classes, and are explained in detail with numerical examples. To implement this approach, the tensor classes PERSPEKTIVE and BASIS are provided, and applied using the functions METRIC and BASES_3D. The complementary graphics class PyPLOT includes basic methods for illustrating the colored objects in two dimensions on the image screen.

- Chapter 6 presents the basics of tensor analysis for the Finite Element Method. Similar to classical approaches [Klingbeil 1966], it initially covers the differential geometry of spatial bodies and then addresses the mechanical basics of the deformation of elastic continua. Tensor representations are then specialized for the approximation approaches of finite elements used to capture geometry and describe displacement fields. These presentations follow classical compendia, such as [Zienkiewicz/Taylor 1989] or the textbook [Meissner/Maurial 2000], but consistently present them in the preferred index notation for matrices and tensors.
  Exemplary matrix relationships are derived for the approximate representation of the geometry of individual elements and for determining stiffness relationships, as well as internal stresses under external loading for numerical analysis. As the focus here is on the principle of matrix methodology, the two-dimensional continua of plate theory are chosen as a relatively straightforward application field. These plate continua are sufficiently complex yet still intuitive, allowing the advantages of the new matrix methods to be demonstrated through typical examples.
- Finally, the application of object-oriented matrix calculus is demonstrated in Sections 7.1 and 7.2 through two typical C++ routines. These routines exemplify in source code the calculation of differential geometric base values, stiffness matrices, and internal forces for various finite plate elements, along with their corresponding numerical results, in a clear and comprehensible program code format.
  These are a bilinear parallelogram element and a linear triangular element, each with abstracted elasticity parameters, enabling the presented calculation results for tension and shear stresses to be easily verified.
  The purpose is to demonstrate how the new matrix classes can create and manage required matrix objects and how matrix arithmetic can be made intuitive and straightforward through overloaded syntax.

In conclusion, the consistent approach pursued here is that tensors, with the benefits and consistency of their theory, also offer a way to efficiently learn numerical methods for designing FEM algorithms. The fundamental understanding of matrix arithmetic and object-oriented software methods is necessary for implementing these in simulation tools in mechanics and engineering. Therefore, these basics are explained in the following chapters.

References

Bathe 1986. Bathe, K.-J.: Finite-Elemente-Methoden. Springer, Berlin (1986)

Borrmann/König/Koch/Beetz (Hrsg.) 2018. Borrmann A.; König M.; Koch Chr.; Beetz J.: Building Information Modeling: Technology Foundations and Industry Practice. Springer, Cham (2018)

Breymann 1997. Breymann, U.: C++ - Eine Einführung. 4. Aufl., Carl Hanser, München (1997)

Clough/Penzien 1975. Clough, R. W.; Penzien, J.: Dynamics of Structures. McGraw-Hill, New York (1975)

Coad/Yourdon 1990. Coad, P.; Yourdon, E.: Object-Oriented Analysis. Prentice Hall, Englewood Cliffs (1990)

Duschek/Hochrainer 1968. Duschek, A.; Hochrainer, A.: Grundzüge der Tensorrechnung in analytischer Darstellung. Springer, Berlin (1968)

Green/Zerna 1954. Green, A. E.; Zerna, W.: Theoretical Elasticity. At The Clarendon Press, Oxford (1954)

Hartmann (Hrsg.) 2000. Hartmann, D.: Objektorientierte Modellierung in Planung und Konstruktion - DFG-Bericht, Wiley-VCH, Weinheim (2000)

Klingbeil 1966. Klingbeil, E.: Tensorrechnung für Ingenieure. Bibliographisches Institut, Mannheim (1966)

Meissner/Maurial 2000. Meißner, U. F.; Maurial, A.: Die Methode der Finiten Elemente - Eine Einführung in die Grundlagen. Springer, Berlin (2000)

Rueppel (Hrsg.) 2007. Rüppel, U.: Vernetzt-kooperative Planungsprozesse im konstruktiven Ingenieurbau : Grundlagen, Methoden, Anwendung und Perspektiven zur vernetzten Ingenieurkooperation. Springer, Berlin (2007)

Rumbaugh/Blaha/Premerlani/Eddy/Lorensen 1991. Rumbaugh, J.; Blaha, M.; Premerlani, W.; Eddy, F.; Lorensen, W,: Object-Oriented Modeling and Design. Prentice Hall, Englewood Cliffs (1991)

Pestel/Leckie 1963. Pestel, E. C.; Leckie, F. A.: Matrix Methods in Elastomechanics. McGraw-Hill, New York (1963)

Zienkiewicz/Taylor 1989. Zienkiewicz, O. C.; Taylor, R. L.: The Finite Element Method. McGraw-Hill, London (1989)

Zurmuehl/Falk 1986. Zurmühl, R.; Falk, S.: Matrizen und ihre Anwendungen. Springer, Berlin (1986)

Chapter 2
Fundamentals of Matrix Calculus

Abstract

In this introductory chapter, the essential fundamentals of classical matrix calculus are compiled, as outlined in [Zurmuehl/Falk 1986] and [Pestel/Leckie 1963]. The primary objective is to acquaint the reader with index notation for multidimensional matrices, which will serve as the basis for all arithmetic operations in the subsequent context. This introduction aims to make the reader comfortable with this notation, which may seem unfamiliar compared to traditional symbolic notation found in related publications. Another important aspect is the generalization of the conventional notation, primarily limited to one- and two-dimensional matrices, which was originally designed for manual computations on traditional media. Therefore adequate formulations for multidimensional matrices are crucial, especially by use of programming languages in the mathematical-scientific field that handle arrays of several dimensions using standard implementations. This introduction to index formulations also prepares the reader for the subsequent acquaintance with covariant and contravariant tensors in the following chapters on tensor analysis and computational tensor/matrix methods.

Traditional matrix calculus harks back to a time when computers did not exist. Consequently, tabular structures were the focus of symbolic matrix notation and processing, not just for numbers but also for variables and functions, intended for manual manipulation. One- and two-dimensional computational schemes, represented on paper, were primarily used for the compact processing of vectors and matrices, especially in the resolution of linear systems of equations. This changed fundamentally with the introduction of programmable computers in the era following Konrad Zuse [Zuse 2019]. Algorithmic programming languages designed for mathematical and scientific purposes have, for a long time now, offered multidimensional arrays as suitable language elements, surpassing the old approach. These arrays can usually be dynamically managed by the operating system during run time. Modern storage media provide exceptional capacities for temporary or permanent access to large data sets, making them ideal for multilevel matrix schemes.

2.1 Notation and Elementary Matrix Algebra

For this new processing of matrices in arrays, which is extensively utilized in engineering nowadays, it is expedient to adjust the classical symbolic notation of matrix cal-

U. F. Meissner, *Tensor Calculus with Object-Oriented Matrices for Numerical Methods in Mechanics and Engineering*, https://doi.org/10.1007/978-3-031-59302-4_2

culus accordingly. The index notation serves as an excellent mathematical counterpart to the arrays used in programming languages. In other words, mathematical operations for matrices can be directly translated into efficient algorithms. This aspect is not new [Duschek/Hochrainer 1968] but has not been as consistently adopted in matrix and tensor calculus in general, because the initial presentation in index notation can be initially challenging due to its apparent complexity. However, this drawback is far outweighed by the clarity of computation rules and the unambiguity in transformations between different reference systems, as will be evident in tensor algebra and analysis later on.

In the following explanation of elementary matrix algebra, symbolic notation and index notation are compared and contrasted in a concise manner. Its understanding requires basic knowledge of matrix calculus, which can be easily acquired from the extensive literature. However, detailed proofs will be largely omitted.

One-dimensional matrices can be represented as ***row vector a*** or ***column vector b*** as follows:

$$\boldsymbol{a} = \begin{pmatrix} a_0 \; a_1 \; \dots \; a_{N_I-1} \end{pmatrix} = a_I \;\; \textit{with column index } I = 0,\dots,N_I - 1 \textit{ and dimension } (1 \times N_I),$$

$$\boldsymbol{b} = \begin{pmatrix} b_0 \\ b_1 \\ \dots \\ b_{N_{K-1}} \end{pmatrix} = b_K \qquad \textit{with row index } K = 0,\dots,N_K - 1 \textit{ and dimension } (N_K \times 1).$$

By multiplying the matrix elements b_K *and* a_I, the ***dyadic product*** results in a rectangular $(N_K \times N_I)$-result matrix $\boldsymbol{D}$

$$\boxed{\boldsymbol{D} = \mathbf{b} \bigotimes \mathbf{a} \;\; \Leftrightarrow \;\; D_{KI} = b_K * a_I = a_I * b_K \quad \textit{with dimension } (N_K \times N_I)}, \qquad (2.1.1)$$

also known as the ***outer product***.

Convention for Matrix Representation

According to the general convention, for a two-dimensional rectangular matrix D_{KI}, the first index K runs over the rows, and the second index I runs over the columns.
(Note: However, in some programming languages, for example, this convention may be also reversed!)

The convention can also be applied to the row and column vectors $\boldsymbol{a}$ and $\boldsymbol{b}$ shown above because they are derived from the set of rectangular matrices. If these vectors are reduced to dimensions $N_I = N_K = 1$, scalar matrix elements like a_0 and b_0 can also be accommodated within this rule.

Therefore both, row vector $\boldsymbol{a}$ and column vector $\boldsymbol{b}$, may be assigned the extended representations of
vector $a_{I0} := a_I$ with dimension $(N_I \times 1)$, respectively
vector $b_{0K} := b_K$ with dimension $(1 \times N_K)$, as well as
scalar $s_0 := s$ with dimension (1)
without changing the values of their elements due to the convention of storage.

In general, it is possible to increase the dimension of a matrix by adding a ***zero index*** through assignment

$$\boxed{D_{KI\mathbf{0}} := D_{K\mathbf{0}I} := D_{\mathbf{0}KI} := D_{KI} \quad \textit{for extended dimensions } N_K,\, N_I,\, 1}, \qquad (2.1.2)$$

and through a reverse assignment to decrease it by one:

$$\boxed{D_{KI} := D_{KI\mathbf{0}} := D_{K\mathbf{0}I} := D_{\mathbf{0}KI} \quad for\ reduced\ dimensions\ N_K,\ N_I}\ , \qquad (2.1.3)$$

which is also of practical importance for the transformation of matrix expressions, as discussed in Section 3.1.

In this context, it is necessary now to clarify the unique notation and distinction of multidimensional matrices in index notation as follows:

Convention for Naming Matrices

- The identity of a matrix is determined by its name (core letter) and the number of its indices. D_{IKL} is therefore different from D_{IK}, and both matrices are unequal: $D_{IKL} \neq D_{IK}$!

- In all mathematical equations, it must be ensured, for consistency of the formulations, that the left side of the ***equality sign '='*** always has the same number of running indices as the right side, and that the same names remain for these indices on both sides: an equalization of D_{IKL} and D_{IKM} would therefore be invalid!

- However, in the case of (2.1.3) a transformation takes place through the assignment, extracting the elements $D_{KIL=0}$ from the three-dimensional matrix D_{KIL} and assigning them to the matrix D_{KI} in terms of reordering. For this purpose, in the following mathematical relationships the ***assignment symbol ':='*** is always used - as it is also common in some programming languages.

Swapping the operands of (2.1.1) in symbolic notation results in the ***scalar product*** with a scalar quantity s of the same dimension N_L for the two vectors through matrix multiplication

$$\boxed{s = \boldsymbol{a} * \boldsymbol{b} \quad \Leftrightarrow \quad s = a_L * b_L = b_L * a_L \quad with\ the\ scalar\ s}\ . \qquad (2.1.4)$$

It is also referred to as the ***inner product*** or simply ***dot product***.

For multiplication, the ***Einstein summation convention*** applies

$$\boxed{s = \sum_{L=0}^{N_L-1} a_L * b_L = a_{\boldsymbol{L}} * b_{\boldsymbol{L}} \quad with\ the\ summation\ index\ \boldsymbol{L} = 0,..,N_L - 1}\ . \qquad (2.1.5)$$

According to this convention, ***the summation sign is always omitted!*** It is worth noting that, due to the index notation, ***the commutativity of the operands is allowed***.

To sharpen the focus on which index the summation is performed, it is rather advisable to initially mark such summation indices in color. Since this option is not available here, they are instead highlighted in bold print.

As can be noticed from this simple example, the following rules apply:

Rules for Index Notation

- In symbolic notation, the arrangement of multiplication operands is not commutative. However, in index notation, these operands can generally be arbitrarily swapped because the computational rule always remains unambiguous. This eliminates the significant restriction of symbolic notation, which severely limits the rearrangement of arithmetic expressions!

- It is also noticeable that the names of the indices can be changed, which - as will be shown later - is of practical importance for the transformation of complex matrix expressions for the sake of clarity.

- The lengths of running indices generally result from the mathematical context of the a priori defined dimensions, and therefore they do not need to be repeated constantly.

- Furthermore, it should be noted that each summation index must always occur in pairs to keep summations unambiguous according to (2.1.5). (Note: The three or more uses of an individually named summation index would lead to errors and are therefore fundamentally excluded!) However, pairs, like other indices, can also be renamed.

- Therefore, it is most advisable to carefully check equations written in index notation.

Alternatively, the scalar product (2.1.4) can also be obtained from matrix D_{KI} according to (2.1.1) through ***contraction*** if it is square. In this case, one gets the ***trace*** of matrix $\boldsymbol{D}$

$$\boxed{s = \boldsymbol{trace}(\boldsymbol{D}) \quad \Leftrightarrow \quad s = D_{LL} \; \textit{as a sum over the main diagonal of } \boldsymbol{D}} \, . \tag{2.1.6}$$

This operation is also called contraction of indices LL which vanish after the summation within the result of s.

The two consecutive operations, (2.1.1) and (2.1.6), for computing the scalar product, however, are not efficient for numerical purposes compared to the direct calculation formula (2.1.4). Firstly, this results in a significant consumption of memory due to the inflation of the matrix $\boldsymbol{D}$, and secondly, this intermediate step leads to many unnecessary multiplications.

In general, the following rules can be derived from this for numerical efficiency:

Rules for Numerical Efficiency

- When chaining matrix operations, it is important to follow a strategy of reducing the size of matrices to avoid inefficient inflation within intermediate steps.

- This is usually achieved by narrowing down matrices in a reasonable sequence of operations.

The scalar product (2.1.4) can be extended to the index notation by introducing additional indices, which aligns with the standard matrix multiplication for rectangular matrices. Similar to the beginning, new matrices from the row and column vectors $\boldsymbol{a}$ and $\boldsymbol{b}$ are created, where the elements are no longer scalars but indexed vectors

$$\boldsymbol{A} = \begin{pmatrix} a_0 \\ a_1 \\ \dots \\ a_{N_{L-1}} \end{pmatrix} = A_{IL} \qquad \textit{with column index } L = 0,..,N_L - 1, \textit{ dimension } (N_I \times N_L),$$

$$\boldsymbol{B} = \left(\boldsymbol{b}_0 \, \boldsymbol{b}_1 \, ... \, \boldsymbol{b}_{N_{L-1}} \right) = B_{LK} \qquad \textit{row index } \; L = 0,..,N_L - 1, \textit{ dimension } (N_L \times N_K).$$

Resulting from the ***matrix multiplication*** yields the two-dimensional result matrix $\boldsymbol{C}$

$$\boxed{\boldsymbol{C} = \boldsymbol{A} * \boldsymbol{B} \;\; \Leftrightarrow \;\; C_{IK} = A_{I\boldsymbol{L}} * B_{\boldsymbol{L}K} \quad \textit{with the rectangular matrix } \boldsymbol{C}} \, . \tag{2.1.7}$$

As evident from the calculation formula, during execution, $N_I * N_K$ scalar products are formed, each with a length of N_L, which are stored in the result matrix C_{IK}.

In the index notation, the ***transposed matrix*** of $\boldsymbol{C}$ can be easily derived by swapping the row and column indices

$$\boxed{\boldsymbol{E} = \boldsymbol{C}^T \;\; \Leftrightarrow \;\; E_{IK} = C_{KI} \quad \textit{with the transposed rectangular matrix } \boldsymbol{C}} \, . \tag{2.1.8}$$

To achieve this, in (2.1.7), one simply has to interchange the indices I and K in order to get

$$\boldsymbol{E} = (\boldsymbol{A} * \boldsymbol{B})^T \;\; \Leftrightarrow \;\; E_{IK} = A_{K\boldsymbol{L}} * B_{\boldsymbol{L}I} \, .$$

This corresponds to

$$\boxed{\boldsymbol{E} = \boldsymbol{B}^T * \boldsymbol{A}^T \;\; \Leftrightarrow \;\; E_{IK} = B_{\boldsymbol{L}I} * A_{K\boldsymbol{L}} \quad \textit{with the rectangular matrix } \boldsymbol{E}} \, . \tag{2.1.9}$$

By transposing the individual matrices in (2.1.7), four different multiplication combinations could be derived in this manner. However, this becomes unnecessary by using the index notation in order to express a ***general matrix multiplication*** as follows:

$$\boxed{C_{IKMN} = A_{I\boldsymbol{L}K} * B_{M\boldsymbol{L}N} \quad \textit{with the four-dimensional result matrix } \boldsymbol{C}}\,, \tag{2.1.10}$$

which is formed by multiplication of two three-dimensional matrices.

To execute all the previously described multiplication variants, one should use the rules from (2.1.2) instead.
First, let's consider the operations for vector matrices:

$$\boxed{D_{0KM0} = b_{0\boldsymbol{0}K} * a_{M\boldsymbol{0}0} \quad \textit{dyadic product (2.1.1) } \boldsymbol{D} = \boldsymbol{b} \otimes \boldsymbol{a}}\,, \tag{2.1.11}$$

$$\boxed{D_{0MK0} = a_{0\boldsymbol{0}M} * b_{K\boldsymbol{0}0} \quad \textit{dyadic product} \quad \boldsymbol{D}^T = \boldsymbol{a} \otimes \boldsymbol{b}}\,,$$

$$\boxed{s_{0000} = a_{0\boldsymbol{L}0} * b_{0\boldsymbol{L}0} \quad \textit{scalar product (2.1.4) } s = \boldsymbol{a} * \boldsymbol{b}}\,. \tag{2.1.12}$$

Next are the matrix multiplications:

$$\boxed{C_{K00M} = A_{K\boldsymbol{L}0} * B_{0\boldsymbol{L}M} \quad \textit{matrix multiplication (2.1.7) } \boldsymbol{C} = \boldsymbol{A} * \boldsymbol{B}}\,, \tag{2.1.13}$$

$$\boxed{C_{0K0M} = A_{0\boldsymbol{L}K} * B_{0\boldsymbol{L}M} \quad \textit{matrix multiplication } \boldsymbol{C} = \boldsymbol{A}^T * \boldsymbol{B}}\,, \tag{2.1.14}$$

$$\boxed{C_{K0M0} = A_{K\boldsymbol{L}0} * B_{M\boldsymbol{L}0} \quad \textit{matrix multiplication } \boldsymbol{C} = \boldsymbol{A} * \boldsymbol{B}^T}\,, \tag{2.1.15}$$

$$\boxed{C_{0KM0} = A_{0\boldsymbol{L}K} * B_{M\boldsymbol{L}0} \quad \textit{matrix multiplication } \boldsymbol{C} = \boldsymbol{A}^T * \boldsymbol{B}^T}\,. \tag{2.1.16}$$

By the interchange of the two indices K and M on the left-hand side, then follows the transposed forms for the result C^T

$$\boxed{C_{0MK0} = A_{0\boldsymbol{L}K} * B_{M\boldsymbol{L}0} \quad \textit{matrix multiplication (2.1.9) } \boldsymbol{C}^T = \boldsymbol{A}^T * \boldsymbol{B}^T}\,. \tag{2.1.17}$$

These general formulations are, of course, independent of numerical efficiency, which must be individually optimized when implementing them in related computation programs, especially for handling large matrices.

The simple ***scalar matrix multiplication*** with a ***factor f*** results in

$$\boxed{C_{IKM} = A_{IKM} * f \quad matrix\ multiplication \quad \mathbf{C} = \mathbf{A} * f}, \tag{2.1.18}$$

where each individual matrix element of $\boldsymbol{C}$ is multiplied by the scalar f.

To conclude the elementary operations, the ***matrix addition*** and ***matrix subtraction*** are listed below

$$\boxed{D_{IKM} = A_{IKM} \pm B_{IKM} \quad matrix\ addition\ and\ subtraction \quad \mathbf{D} = \mathbf{A} \pm \mathbf{B}}. \tag{2.1.19}$$

2.2 Special Matrices

In the following, a brief introduction to the handling of special matrices is provided.

First, let's consider the ***symmetry of matrices***. Matrices are symmetric if swapping two index positions does not change the value of the matrix

$$\boxed{A_{IK} = A_{KI} \quad symmetric\ in\ I\ and\ K \quad \mathbf{A} = \mathbf{A}^T}. \tag{2.2.1}$$

A typical representative of this category is the ***identity matrix I***, also known as the ***Kronecker delta*** in index notation, or simply as the $\boldsymbol{\delta - matrix}$

$$\boxed{\delta_{IK} = \begin{Bmatrix} +1, \ if\ I = K \\ 0, \ if\ I \neq K \end{Bmatrix} = \boldsymbol{I}}. \tag{2.2.2}$$

In matrix form, for a (2×2)- or (3×3)-matrix, it is represented as

$$\delta_{IK} = \begin{pmatrix} 1\,0 \\ 0\,1 \end{pmatrix} \qquad \delta_{IK} = \begin{pmatrix} 1\,0\,0 \\ 0\,1\,0 \\ 0\,0\,1 \end{pmatrix}$$

with the symmetry $\delta_{IK} = \delta_{KI}$.

In general, it satisfies the relationship

$$spur(\boldsymbol{I} * \boldsymbol{I}^T) = \delta_{K\boldsymbol{I}} * \delta_{K\boldsymbol{I}} = \delta_{\boldsymbol{KK}} = n \ with\ the\ dimension\ N_I = N_K = n\ .$$

The δ-matrix plays an important role because it allows for the ***exchange of indices*** in expressions like

$$A_{pqL} = \delta_{p\boldsymbol{I}} * \delta_{q\boldsymbol{K}} * A_{\boldsymbol{IK}L}$$

to be redrafted.

Next, the ***antisymmetry of matrices*** is considered. Matrices are antisymmetric or skew-symmetric if swapping two index positions changes the sign of the matrix

$$\boxed{A_{IK} = -A_{KI} \quad antisymmetric\ in\ I\ und\ K \quad \mathbf{A} = -\mathbf{A}^T}\,, \tag{2.2.3}$$

and the diagonal elements $A_{II} = 0$ (no summation over I) are zero.

A typical representative of this category is the ***Levi-Civita symbol***, also known as the ***permutation symbol*** resp. in index notation in short as the ***e-matrix***

$$\boxed{\mathrm{e}_{IKL...} = \begin{cases} +1, for\ (I,K,L...)\ as\ an\ even\ permutation\ of\ (0,1,2...)\,, \\ -1, for\ (I,K,L...)\ as\ an\ odd\ permutation\ of\ (2,1,0...)\,, \\ 0,\ if\ at\ least\ two\ indices\ are\ the\ same. \end{cases}} \tag{2.2.4}$$

In compact form, for a $(2\times2)-$ or $(3\times3)-$matrix, it is represented as

$$e_{IK} = \begin{pmatrix} 0 & 1 \\ -1 & 0 \end{pmatrix} \quad e_{IK0} = \begin{pmatrix} 0 & 0 & 0 \\ 0 & 0 & 1 \\ 0 & -1 & 0 \end{pmatrix} \quad e_{IK1} = \begin{pmatrix} 0 & 0 & -1 \\ 0 & 0 & 0 \\ 1 & 0 & 0 \end{pmatrix} \quad e_{IK2} = \begin{pmatrix} 0 & 1 & 0 \\ -1 & 0 & 0 \\ 0 & 0 & 0 \end{pmatrix}$$

with antisymmetries $e_{IK} = -e_{KI}$ resp. $e_{IKL} = -e_{KIL}$, $e_{IKL} = -e_{ILK}$, $e_{IKL} = -e_{LKI}$.

The ***e-matrix*** (2.2.4) can also be defined as a determinant of the mth order using the δ-matrix (2.2.2)

$$\boxed{e_{IKL...P} = \begin{vmatrix} \delta_{0I} & \delta_{0K} & \delta_{0L} & ..\ \delta_{0P} \\ \delta_{1I} & \delta_{1K} & \delta_{1L} & ..\ \delta_{1P} \\ \delta_{2I} & \delta_{2K} & \delta_{2L} & ..\ \delta_{2P} \\ .. & .. & .. & ..\ .. \\ \delta_{m-1I} & \delta_{m-1K} & \delta_{m-1L} & ..\ \delta_{m-1P} \end{vmatrix} \quad N_I = N_K = N_L = N_P = m}\,. \tag{2.2.5}$$

For a $(2\times2)-$ or a $(3\times3)-$matrix, it can be represented compactly as

$$e_{IK} = \begin{vmatrix} \delta_{SI}\ \delta_{SK} \end{vmatrix}\ with\ N_S = 2\ \ resp.\ \ e_{IKL} = \begin{vmatrix} \delta_{SI}\ \delta_{SK}\ \delta_{SL} \end{vmatrix}\ with\ N_S = 3.$$

This leads to the ***Laplace expansion theorem*** as a δ-matrix of the 6th order

$$\delta_{IKLMNP} = e_{IKL} * e_{MNP} = \begin{vmatrix} \delta_{SI} \\ \delta_{SK} \\ \delta_{SL} \end{vmatrix} * \begin{vmatrix} \delta_{SM} & \delta_{SN} & \delta_{SP} \end{vmatrix} \quad with\, N_S = 3$$

$$= \begin{vmatrix} \delta_{IM} & \delta_{IN} & \delta_{IP} \\ \delta_{KM} & \delta_{KN} & \delta_{KP} \\ \delta_{LM} & \delta_{LN} & \delta_{LP} \end{vmatrix} = \delta_{IM} * \begin{vmatrix} \delta_{KN} & \delta_{KP} \\ \delta_{LN} & \delta_{LP} \end{vmatrix} - \delta_{IN} * \begin{vmatrix} \delta_{KM} & \delta_{KP} \\ \delta_{LM} & \delta_{LP} \end{vmatrix} + \delta_{IP} * \begin{vmatrix} \delta_{KM} & \delta_{KN} \\ \delta_{LM} & \delta_{LN} \end{vmatrix} . \tag{2.2.6}$$

From (2.2.5), the ***cross product*** of two vectors $\boldsymbol{b}$ and $\boldsymbol{c}$ is derived as follows

$$\boxed{v_I = e_{I\boldsymbol{KL}} * b_{\boldsymbol{K}} * c_{\boldsymbol{L}} = \begin{vmatrix} \delta_{SI} & b_S & \delta_{S\boldsymbol{L}} \end{vmatrix} * c_{\boldsymbol{L}} = \begin{vmatrix} \delta_{SI} & b_S & c_S \end{vmatrix} \Leftrightarrow \boldsymbol{v} = \boldsymbol{b} \times \boldsymbol{c}} , \tag{2.2.7}$$

and by scalar multiplication with a vector $\boldsymbol{a_I}$ the ***scalar triple product***

$$\boxed{s = a_{\boldsymbol{I}} * e_{\boldsymbol{IKL}} * b_{\boldsymbol{K}} * c_{\boldsymbol{L}} \quad \Leftrightarrow \quad s = \boldsymbol{a} * (\boldsymbol{b} \times \boldsymbol{c})} \tag{2.2.8}$$

is obtained.

Multiply (2.2.5) with the columns of a correspondingly dimensioned matrix $A_{IKL..P}$

$$e_{\boldsymbol{IKL..P}} * A_{\boldsymbol{I}0} * A_{\boldsymbol{K}1} * A_{\boldsymbol{L}2} * .. * A_{\boldsymbol{P}m-1}$$
$$= \begin{vmatrix} \delta_{\boldsymbol{I}S} & \delta_{\boldsymbol{K}S} & \delta_{\boldsymbol{L}S} & .. & \delta_{\boldsymbol{P}S} \end{vmatrix} * A_{\boldsymbol{I}0} * A_{\boldsymbol{K}1} * A_{\boldsymbol{L}2} * .. * A_{\boldsymbol{P}m-1}$$
$$= \begin{vmatrix} A_{S0} & A_{S1} & A_{S2} & .. & A_{Sm-1} \end{vmatrix} = det(\boldsymbol{A}) \quad with\, N_S = m$$

yields in the ***determinant***

$$\boxed{det(\boldsymbol{A}) = e_{\boldsymbol{IKL..P}} * A_{\boldsymbol{I}0} * A_{\boldsymbol{K}1} * A_{\boldsymbol{L}2} * .. * A_{\boldsymbol{P}m-1}} . \tag{2.2.9}$$

For $(2 \times 2)-$matrices, it follows:

$$det(\boldsymbol{A}) = e_{\boldsymbol{IK}} * A_{\boldsymbol{I}0} * A_{\boldsymbol{K}1} = e_{\boldsymbol{IK}} * e_{01} * A_{\boldsymbol{I}0} * A_{\boldsymbol{K}1} .$$

By adding its equivalent with the permutation e_{10}

$$det(\boldsymbol{A}) = e_{\boldsymbol{IK}} * e_{10} * A_{\boldsymbol{I}1} * A_{\boldsymbol{K}0}$$

the **(2×2)-*determinant by matrix multiplication***

$$\boxed{det(\boldsymbol{A}) = 1/2 * e_{\boldsymbol{IK}} * e_{\boldsymbol{MN}} * A_{\boldsymbol{IM}} * A_{\boldsymbol{KN}}} \tag{2.2.10}$$

is obtained.
In a similar manner, the **(3×3)-*determinant by matrix multiplication***

$$\boxed{det(\boldsymbol{A}) = 1/6 * e_{\boldsymbol{IKL}} * e_{\boldsymbol{MNP}} * A_{\boldsymbol{IM}} * A_{\boldsymbol{KN}} * A_{\boldsymbol{LP}}} \tag{2.2.11}$$

may be achieved, where the prefactor arises from the 6-fold permutation of the e-matrix elements.

If the δ-matrix δ_{SI} in equation (2.2.6) is replaced with a matrix $\boldsymbol{A}_{SI}$, and δ_{SM} with $\boldsymbol{B}_{SM}$, the well-known ***product of determinants***

$$\boxed{det(A_{\boldsymbol{SI}} * B_{\boldsymbol{SM}}) = det(A_{\boldsymbol{SI}}) * det(B_{\boldsymbol{SM}}) \quad \Leftrightarrow \quad det(\boldsymbol{A}^{\boldsymbol{T}} * \boldsymbol{B}) = det(\boldsymbol{A}^T) * det(\boldsymbol{B})} \tag{2.2.12}$$

is achieved.

Because the value of the determinant (2.2.5) does not change when rows and columns are transposed

$$det(\boldsymbol{A}^{\boldsymbol{T}}) = det(\boldsymbol{A}),$$

it also holds that

$$det(A_{\boldsymbol{IS}} * B_{\boldsymbol{SM}}) = det(\boldsymbol{A}) * det(\boldsymbol{B}) = det(\boldsymbol{A}^T) * det(\boldsymbol{B}),$$

similarly corresponding to

$$det(\boldsymbol{A}^k) = (det(\boldsymbol{A}))^k.$$

2.3 Variational Methods and Linear Systems of Equations

In numerical mechanics and engineering applications, linear systems of equations and algorithms for their numerical solutions play a central role. Therefore, in the following, the essential fundamentals are once again summarized - without claiming completeness - to further clarify and explain the formulations in index notation. The generally accepted terminology of the Finite Element Method (FEM), commonly used for the mathematical description of discretized computational systems, will be used [Zienkiewicz/Taylor 1989], [Bathe 1986], or [Meissner/Maurial 2000].

A ***quadratic form*** is defined as a matrix formulation, whose left-hand side Q is a scalar quantity

$$\boxed{Q = x_I * A_{IK} * x_K \quad \Leftrightarrow \quad Q = \boldsymbol{x}^T * \boldsymbol{A} * \boldsymbol{x}}\,, \tag{2.3.1}$$

only when the ***coefficient matrix*** is symmetric with $A_{IK} = A_{KI}$ resp. $\boldsymbol{A} = \boldsymbol{A}^T$.

If for arbitrary real values of $x_I \neq 0$, the numerical value of Q is always positive, then the form is called positive definite. Conversely, if Q can also take the value 0, it is called positive semi-definite. On the other hand, the form is negative definite if Q is always negative.

With respect to the quadratic form also the coefficient matrix $\boldsymbol{A}$ is characterized as definite as follows:

A_{IK} is ***positive definite***, if Q > 0,
positive semi-definite, if $Q \geq 0$,
negative definite, if Q < 0,
negative semi-definite, if $Q \leq 0$,
actually definite or ***regular***, if $Q \neq 0$
and
singular, if Q = 0.

These properties depend on the determinant of the matrix A_{IK}. Because for a singular matrix

$$det(A_{IK}) = 0 \quad \Leftrightarrow \quad det(\boldsymbol{A}) = \boldsymbol{0}$$

there exist non-trivial solutions $x_I \neq 0$ that make the value of the quadratic form Q vanish. This can be recognized from (2.3.1) using the determinant product

$$Q = det(x_I * A_{IK} * x_K) = det(A_{IK}) * det(x_I * x_K) = 0\,.$$

In the Finite Element Method (shortly FEM resp. FE-Method), quadratic forms play a prominent role in formulating numerical calculation methods for deformable structures, as the (internally stored) ***potential energy of internal forces*** of an elastic system is given by

$$\Pi^{(i)} = 1/2 * V_I * K_{IK} * V_K \quad \Leftrightarrow \quad \Pi^{(i)} = 1/2 * \boldsymbol{V}^T * \boldsymbol{K} * \boldsymbol{V}$$

where K_{IK} represents the ***symmetric global stiffness matrix*** of a complex mechanical/static system, and V_K is the ***displacement vector*** with the kinematic degrees of freedom of the discrete element nodes.

The ***potential energy of external forces*** is referred to as the displacements work

$$\Pi^{(a)} = -V_I * F_I \quad \Leftrightarrow \quad \Pi^{(a)} = -\boldsymbol{V}^T * \boldsymbol{F}$$

which represents the work done by the applied force quantities F_I during the deformation of the system along the displacements V_I .

The sum of both components

$$\Pi = \Pi^{(i)} + \Pi^{(a)}$$

results in the ***(total) potential energy*** of the system

$$\boxed{\Pi = 1/2 * V_I * K_{IK} * V_K - V_I * F_I \quad \Leftrightarrow \quad \Pi = 1/2 * \boldsymbol{V}^T * \boldsymbol{K} * \boldsymbol{V} - \boldsymbol{V}^T * \boldsymbol{F}} \,. \tag{2.3.2}$$

Similar relationships also arise in many other areas of applied physics, such as fluid mechanics, electrical engineering, acoustics, and optics, when appropriate potentials are present. Extensive relevant literature refers to these fields.

The formulation of Π represents a ***variational functional*** in which the individual vector elements V_I are variables of functions whose structure and coordinates are invisable deposited.

For the variation of the functional, the rules of partial differentiation apply

$$\delta\Pi = \partial\Pi/\partial(V_1) * \delta(V_1) + \partial\Pi/\partial(V_2) * \delta(V_2) + ...$$

like for the complete differential of a function $f = f(\Theta_1, \Theta_2, ...)$

$$df = \partial f/\partial(\Theta_1) * d(\Theta_1) + \partial f/\partial(\Theta_2) * d(\Theta_2) + ... $$

The variations $\delta(V_I)$ are also assigned the same meaning as the differentials $d(\Theta_i)$ in differential calculus.

Requirements for variations $\delta(V_I)$ are:
- they must be sufficiently small,
- they should be arbitrary, and
- they must also be compatible with prescribed values of V_I !

Thus, they have the same meaning as the ***virtual displacements*** introduced as "variations" of the kinematic quantities (displacements or rotations) in the corresponding principle of mechanics.

Regarding the nomenclature of partial derivatives and differentiation with respect to specific vector components, the following commonly used notations are adopted for index notation:

- ***partial derivatives*** are formulated using the abbreviated ***comma notation***

$$\boxed{\partial\Pi/\partial(V_I) = \Pi,_I \quad resp. \quad \partial f/\partial(\Theta_\alpha) = f,_\alpha} \tag{2.3.3}$$

and

- ***Scalars or matrices are differentiated with respect to a vector*** by differentiating their functional elements individually with respect to the variable components of the vector and by storing the results column-wise.

Examples:
the ***first-order variation of potential energy***

$$\boxed{\delta\Pi = \left(\, \partial\Pi/\partial(V_1) \;\; \partial\Pi/\partial(V_2) \; \ldots \right) * \delta V_I = \Pi,_I * \delta V_I} \; , \tag{2.3.4}$$

the ***second-order variation of potential energy***

$$\boxed{\delta^2\Pi = \left(\, \partial(\Pi,_I)/\partial(V_1) \;\; \partial(\Pi,_I)/\partial(V_2) \; \ldots \right) * \delta V_I * \delta V_K = \Pi,_{IK} * \delta V_I * \delta V_K} \; . \tag{2.3.5}$$

To determine the potential energy (2.3.2) stored in the mechanically/statically system due to deformation, its minimum is sought according to the rules of differential calculus.

For this purpose equation (2.3.4) provides the necessary condition

$$\delta\Pi = 1/2 * (\delta V_I * K_{IK} * V_K + V_I * K_{IK} * \delta V_K) - \delta V_I * F_I = 0.$$

With the ***symmetry*** of $K_{IK} = K_{KI}$ and by interchanging the indices I and K in the second term, the ***necessary condition for the existence of a minimum*** of potential energy is obtained

$$\boxed{\delta\Pi = \delta V_I * (K_{IK} * V_K - F_I) = 0 \quad \Leftrightarrow \quad \delta\Pi = \delta \boldsymbol{V}^T * (\boldsymbol{K} * \boldsymbol{V} - \boldsymbol{F}) = 0} \; , \tag{2.3.6}$$

which corresponds to the ***principle of virtual displacements*** in the theory of mechanics, where the first-order variation of the potential energy functional must be zero.

Because the variations δV_I are arbitrary, it directly follows from this condition, the ***linear equation system***

$$\boxed{K_{IK} * V_K = F_I \quad \Leftrightarrow \quad \boldsymbol{K} * \boldsymbol{V} = \boldsymbol{F}} \; . \tag{2.3.7}$$

From the second-order variation (2.3.5), the ***sufficient condition for the existence of the minimum*** is obtained

$$\boxed{\delta^2\Pi = \delta V_I * K_{IK} * \delta V_K > 0 \quad \Leftrightarrow \quad \delta^2\Pi = \delta \boldsymbol{V}^T * \boldsymbol{K} * \delta \boldsymbol{V} > 0} \tag{2.3.8}$$

with the open question of whether this condition is met or not.

In general, it can be recognized that this condition is not satisfied because the determinant of K_{IK} is initially singular:

$$\delta^2\Pi = det(K_{IK}) * det(\delta V_I * \delta V_K) = 0 \, .$$

So, as already mentioned, there exist non-trivial solutions for $V_I \neq 0$ that let $\delta^2\Pi$ vanish and do not generate any potential energy of internal forces.

The reason for this lies in the nature of the underlying mechanical boundary value problem. As long as the essential boundary conditions resp. prescribed values of V_I are not incorporated in (2.3.6), rigid body motions can shift the mechanical system arbitrarily without triggering internal forces.

In order to incorporate such geometrical boundary conditions, the displacement vector V_I is split into two complementary parts using a Boolean matrix S_{IL}. The part $\bar{V}_I$ contains the prescribed essential boundary values, and the part $\hat{V}_I$ includes only the unknown unrestricted values

$$\boxed{V_I = S_{IL} * \hat{V}_L + \bar{V}_I \quad with \quad \delta V_I = S_{IL} * \delta\hat{V}_L \quad because \quad \delta\bar{V}_I \equiv 0} . \qquad (2.3.9)$$

This split is achieved using the Boolean matrix S_{IL} describing the support conditions of the mechanical/statical system, which filters out the boundary constraints of the system by using zeros and ones, thus reducing the number of kinematic degrees of freedom in the equation system.
For example:

$$V_I = \begin{pmatrix} V_0 \\ V_1 \\ V_2 \\ \dots \end{pmatrix} = \begin{pmatrix} 1 & 0 & \dots \\ 0 & 0 & \dots \\ 0 & 1 & \dots \\ \dots & \dots & \dots \end{pmatrix} * \begin{pmatrix} \hat{V}_0 \\ \hat{V}_2 \\ \dots \end{pmatrix} + \begin{pmatrix} 0 \\ \bar{V}_1 \\ 0 \\ \dots \end{pmatrix} = S_{IL} * \hat{V}_L + \bar{V}_I \, .$$

For the force quantities F_I on the right-hand side of (2.3.7), a corresponding complementary splitting of the statical boundary conditions is made into two parts using

$$\boxed{F_I = \bar{F}_I + \hat{F}_I} . \qquad (2.3.10)$$

The first part $\bar{F}_I$ of the load vector contains the prescribed external load values, and the second part $\hat{F}_I$ includes the as-yet-unknown support forces, which result from the solution of the equation system as natural boundary conditions after the numerical computation.

Substituting (2.3.9) and (2.3.10) into (2.3.6), one obtains the first variation

$$\delta\hat{\Pi} = \delta\hat{V}_L * (\bar{K}_{LM} * \hat{V}_M - \bar{R}_L) = 0 \quad \Leftrightarrow \quad \delta\hat{\Pi} = \delta\hat{\boldsymbol{V}}^T * (\bar{\boldsymbol{K}} * \hat{\boldsymbol{V}} - \bar{\boldsymbol{R}}) = 0 \, ,$$

which, due to the arbitrariness of $\delta\hat{V}_L$, reduces to the ***reduced equation system***

$$\boxed{\bar{K}_{LM} * \hat{V}_M = \bar{R}_L \quad \Leftrightarrow \quad \bar{\boldsymbol{K}} * \hat{\boldsymbol{V}} = \bar{\boldsymbol{R}}} \qquad (2.3.11)$$

with the regular stiffness matrix given by

$$\bar{K}_{LM} = S_{IL} * K_{IK} * S_{KM} \quad \Leftrightarrow \quad \bar{\boldsymbol{K}} = \boldsymbol{S}^T * \boldsymbol{K} * \boldsymbol{S}$$

and due to

$$S_{IL} * \hat{F}_I \equiv 0 \quad \Leftrightarrow \quad \boldsymbol{S}^T * \hat{\boldsymbol{F}} \equiv \boldsymbol{0}$$

leads to the reduced right-hand side

$$\bar{R}_L = S_{IL} * (K_{IM} * \bar{V}_M - \bar{F}_I) \quad \Leftrightarrow \quad \bar{R} = \boldsymbol{S}^T * (\boldsymbol{K} * \bar{\boldsymbol{V}} - \bar{\boldsymbol{F}}) \,.$$

The determinant of the reduced coefficient matrix is now positive

$$det(\bar{K}_{LM}) > 0 \quad \Leftrightarrow \quad det(\bar{\boldsymbol{K}}) > 0 \,,$$

if all rigid body motions are incorporated through the 'elimination of rows and columns' using (2.3.9). Consequently, according to (2.3.8), the presence of the minimum of potential energy is hereby sufficiently demonstrated.

After solving the system of equations (2.3.11), the unknown force quantities $\hat{F}_I$ can be calculated. To do this, substitute (2.3.9) and (2.3.10) into the relationship (2.3.7), obtaining the ***support reactions*** as follows:

$$\boxed{\hat{F}_I = K_{IK} * (S_{KL} * \hat{V}_L + \bar{V}_K) - \bar{F}_I \quad \Leftrightarrow \quad \hat{\boldsymbol{F}} = \boldsymbol{K} * (\boldsymbol{S} * \hat{\boldsymbol{V}} + \bar{\boldsymbol{V}}) - \bar{\boldsymbol{F}}} \,. \tag{2.3.12}$$

By conclusion of these considerations, the question arises as to how large the minimum amount of the potential energy is after deformation of the mechanical/static system. For this purpose, (2.3.6) is used. Because this relationship is valid for arbitrary variations δV_I in general, it therefore holds also for the specific calculated solution V_I. Specifically, the relation

$$V_I * (K_{IK} * V_K - F_I) \equiv 0$$

follows.
Re-substituting this relationship into (2.3.2), the calculated ***value of the minimum potential energy***

$$\boxed{\Pi^{(min)} = -1/2 * V_I * K_{IK} * V_K \quad \Leftrightarrow \quad \Pi^{(min)} = -1/2 * \boldsymbol{V}^T * \boldsymbol{K} * \boldsymbol{V}} \tag{2.3.13}$$

is obtained.

As already mentioned earlier, the presented approach has an exemplary character in cases where a corresponding field potential exists. It should be noted that in many practical applications of applied physics, such as in the numerical modeling of Navier-Stokes flows, this is often not valid. However, in such cases, the numerical approach typically begins with a similar initial formulation as (2.3.6), often with additional constraints. For the numerical approximation of the underlying differential equations, then instead of the presented Rayleigh-Ritz variational method the more general approach according to Bubnov-Galerkin is followed as a modified method, which minimizes the residuals between exact and approximate solutions sufficiently, without having an exact proof for the minimum (2.3.13) as demonstrated.

For the ***solution of linear equation systems*** in general

$$\boxed{\mathrm{A}_{IK} * x_K = r_I \quad \Leftrightarrow \quad \boldsymbol{A} * \boldsymbol{x} = \boldsymbol{r}}\,, \tag{2.3.14}$$

two approaches are commonly used. On one hand, there are traditionally ***direct decomposition methods*** available, and on the other hand, modern ***iterative methods with preconditioning*** of the coefficient matrix $\boldsymbol{A}$ are highly efficient. The latter, especially when executed on parallel computers, significantly enhance computational performance (speed-up) as the number of required computational operations can be drastically reduced. For explanations and understanding of the corresponding algorithms, the relevant literature of this topic is referred. The first category is extensively covered in [Zurmuehl/Falk 1986] and further explained for programming in [Meissner/Maurial 2000]. The second category is detailed with appropriate methods in [Hackbusch 1991] and discussed in terms of efficiency criteria. Furthermore, it was scientifically investigated i.a. by [Laemmer 1996] for Civil and Structural Engineering applications, especially for its implementation on parallel computer systems.

Decomposition methods are fundamentally based on decomposing the (n×n)-coefficient matrix $\boldsymbol{A}$ into a ***lower triangular matrix L*** and an ***upper triangular matrix U***

$$\boxed{\mathrm{A}_{IK} = L_{IL} * U_{LK} \quad \Leftrightarrow \quad \boldsymbol{A} = \boldsymbol{L} * \boldsymbol{U}}\,, \tag{2.3.15}$$

where the solution $\boldsymbol{x}$ of the equation system is obtained in two separate steps. First, the intermediate solution $\boldsymbol{y}$ is achieved through ***forward substitution*** in $\boldsymbol{L}$

$$L_{IK} * y_K = r_I \quad \Leftrightarrow \quad \boldsymbol{L} * \boldsymbol{y} = \boldsymbol{r},$$

and then followed by determination of $\boldsymbol{x}$ by ***backward substitution*** in $\boldsymbol{U}$

$$U_{IK} * x_K = y_I \quad \Leftrightarrow \quad \boldsymbol{U} * \boldsymbol{x} = \boldsymbol{y}.$$

This triangular decomposition scheme is illustrated as follows:

$$A_{IK} = L_{IL} * U_{LK} \qquad \begin{pmatrix} U_{00} & U_{01} & U_{02} & U_{03} & U_{04} & \dots \\ & U_{11} & U_{12} & U_{13} & U_{14} & \dots \\ & & U_{22} & U_{23} & U_{24} & \dots \\ & & & U_{33} & U_{34} & \dots \\ & & & & U_{44} & \dots \\ & & & & & \dots \end{pmatrix} = U_{LK}$$

$$L_{IL} = \begin{pmatrix} L_{00} & & & & & \\ L_{10} & L_{11} & & & & \\ L_{20} & L_{21} & L_{22} & & & \\ L_{30} & L_{31} & L_{32} & L_{33} & & \\ L_{40} & L_{41} & L_{42} & L_{43} & L_{44} & \\ \dots & \dots & \dots & \dots & \dots & \dots \end{pmatrix} \begin{pmatrix} A_{00} & A_{01} & A_{02} & A_{03} & A_{04} & \dots \\ A_{10} & A_{11} & A_{12} & A_{13} & A_{14} & \dots \\ A_{20} & A_{21} & A_{22} & A_{23} & A_{24} & \dots \\ A_{30} & A_{31} & A_{32} & A_{33} & A_{34} & \dots \\ A_{40} & A_{41} & A_{42} & A_{43} & A_{44} & \dots \\ \dots & \dots & \dots & \dots & \dots & \dots \end{pmatrix} = A_{IK}.$$

For ***symmetric coefficient matrices*** as in (2.3.7), two decomposition methods are suitable:

a) The ***Cholesky factorization algorithm***, if A_{IK} is positive-definite.
This method decomposes $\boldsymbol{A}$ into two dual upper and lower matrices. The lower matrix $\boldsymbol{L}$ and the upper matrix $\boldsymbol{U}$ are transposed, and represented by a normalized triangular matrix $\boldsymbol{G}$ with ones on its diagonal as well as by a diagonal matrix $\boldsymbol{D}$ generated by square roots of its elements:

$$L_{KL} = U_{LK} = D_{LM}^{1/2} * G_{MK} \quad \Leftrightarrow \quad \boldsymbol{L^T} = \boldsymbol{U} = \boldsymbol{D}^{1/2} * \boldsymbol{G}$$

where

$$det(A_{IK}) = \boldsymbol{\Pi}_{i=0}^{n-1} D_{ii}^2 \quad \Leftrightarrow \quad det\boldsymbol{A} = \boldsymbol{\Pi}_{i=0}^{n-1} D_{ii}^2 \,.$$

Due to the square root this algorithm helps to a certain amount mitigating rounding errors.

b) The ***Gauss elimination algorithm***, if A_{IK} is actually definite.
This method also decomposes $\boldsymbol{A}$ into upper and lower matrices. The lower matrix $\boldsymbol{L}$ and the upper matrix $\boldsymbol{U}$ are split using a normalized triangular matrix $\boldsymbol{G}$ with ones on the main diagonal and a diagonal matrix $\boldsymbol{D}$, with its elements formed by divisions:

$$L_{IL} = G_{LI} \quad U_{LK} = D_{LM} * G_{MK} \quad \Leftrightarrow \quad \boldsymbol{L} = \boldsymbol{G^T} \quad \boldsymbol{U} = \boldsymbol{D} * \boldsymbol{G}$$

where

$$det(A_{IK}) = \boldsymbol{\Pi}_{i=0}^{n-1} D_{ii} \quad \Leftrightarrow \quad det\boldsymbol{A} = \boldsymbol{\Pi}_{i=0}^{n-1} D_{ii} \,.$$

In the case of negative definiteness, divisions by zero may occur and must be handled by pivoting (swapping of rows and columns).

The numerical effort for the decomposition of the lower triangular matrix is approximately the same for both methods, involving $1/3 * n^3 + \mathbf{O}(n^2)$ arithmetic operations with $1/6 * n^3$ multiplications, $1/2 * n^2$ divisions (resp. multiplications by reciprocals), and n square root operations resp. divisions on the main diagonal.

The ***Gauss elimination algorithm*** with eventual pivoting is also suitable ***for non-symmetric coefficient matrices***. However, it requires approximately twice the numerical effort, as both the normalized lower matrix $\boldsymbol{L}$ and the upper matrix $\boldsymbol{U}$ must be determined separately. Due to $det(\boldsymbol{L}) = 1$ the determinant of the coefficient matrix results simply from diagonal elements of the upper matrix

$$det(A_{IK}) = \Pi_{i=0}^{n-1} U_{ii} \quad \Leftrightarrow \quad det\boldsymbol{A} = \Pi_{i=0}^{n-1} U_{ii} \; .$$

For backward substitution to determine the solutions $\boldsymbol{x}$ following (2.3.15), all three methods involve a numerical effort of $1/2 * n^2 + \mathbf{O}(n)$ arithmetic operations. The main effort of the direct decomposition methods is therefore in generating the staggered triangular matrices, which is of order $\mathbf{O}(n^3)$ arithmetic operations. Iterative methods excel in this aspect and significantly reduce the numerical effort [Laemmer 1996] for very large equation systems through efficient preconditioning and parallelizing algorithms [Laemmer/Meissner/Ruben 2003].

With the described decomposition methods, the ***inverse matrix*** Z_{KL} of a regular coefficient matrix A_{IK} may also be computed, using the identity matrix δ_{IL} when n columns are processed simultaneously on the right-hand side of (2.3.14)

$$\boxed{\mathrm{A}_{IK} * Z_{KL} = \delta_{IL} \quad \Leftrightarrow \quad \boldsymbol{A} * \boldsymbol{Z} = \boldsymbol{I} \quad mit \quad \boldsymbol{Z} = \boldsymbol{A}^{-1}} \; . \tag{2.3.16}$$

References

Bathe 1986. Bathe, K.-J.: Finite-Elemente-Methoden. Springer, Berlin (1986)

Duschek/Hochrainer 1968. Duschek, A.; Hochrainer, A.: Grundzüge der Tensorrechnung in analytischer Darstellung. Springer, Berlin (1968)

Hackbusch 1991. Hackbusch, W.: Iterative Lösung großer schwachbesetzter Gleichungssysteme. Teubner, Stuttgart - Studienbücher Mathematik (1991)

Laemmer 1996. Lämmer, L.: Parallelisierung von Anwendungen der Finite-Element-Methode im Bauingenieurwesen. Habilitationsschrift TU Darmstadt (1996)

Laemmer/Meissner/Ruben 2003. Lämmer L., Meißner U., Ruben J.: Finite Element Modelling of Soil-Structure Systems in Workstation Clusters. Proceedings Second MIT Conference on Computational Fluid and Solid Mechanics, Vol. 2, P. 2312, Elsevier Cambridge (2003)

Meissner/Maurial 2000. Meißner, U. F.; Maurial, A.: Die Methode der Finiten Elemente - Eine Einführung in die Grundlagen. Springer, Berlin (2000)

Pestel/Leckie 1963. Pestel, E. C.; Leckie, F. A.: Matrix Methods in Elastomechanics. McGraw-Hill, New York (1963)

Zienkiewicz/Taylor 1989. Zienkiewicz, O. C.; Taylor, R. L.: The Finite Element Method. McGraw-Hill, London (1989)

Zurmuehl/Falk 1986. Zurmühl, R.; Falk, S.: Matrizen und ihre Anwendungen. Springer, Berlin (1986)

***

Internet Links

Zuse 2019. Zuse, H.: Konrad Zuse Multimedia Show (2018/19)
http://www.horst-zuse.homepage.t-online.de/_wp_generated/zuse-broschuere.html

Chapter 3
Object-Oriented Matrices

Abstract

In this chapter on object-oriented matrices, we first focus on the structure and storage of multidimensional matrices, which are also mapped by applications on digital von Neumann computers into one-dimensional physical main memory. The section on hypermatrices explains this context and presents different representations for multidimensional matrices with the corresponding mathematical mapping rules. The representation of matrices is followed by the exemplary implementation of multidimensional matrices and vectors in object-oriented matrix classes, like MATRIZEM, MATRIX and VEKTOR, using the programming language C++. The primary goal is to demonstrate the organization of matrix objects in suitable class structures according to the object-oriented principles of encapsulation, inheritance, and communication interfaces of public functions. Readers may extend their knowledge of object orientation in C++ by studying relevant literature, such as [Rumbaugh/Blaha/Premerlani/Eddy/Lorensen 1991] on object-oriented modeling or like [Breymann 1997] on C++, as well as conduct research on the Internet. In addition to generating and managing static and dynamic matrix objects, it is particularly important to recognize the advantages of operator overloading in the presented approach. This allows for an intuitive and transparent handling of the new matrix objects for arithmetic and functional operations. These aspects will be demonstrated in Chapter 7 through two typical FEM-application algorithms implemented using tensor and matrix objects.

After the previous chapters have explained the basics of nomenclature and computational rules, the following sections will focus now on the handling of multidimensional matrices and their algorithmic processing for numerical tasks.

For this purpose, a consistent convention for the representation and storage of multidimensional matrices in computational programs is required. This convention varies across different programming languages. For instance, in C, C++, and Java, arrays are stored row-wise, while in FORTRAN, column-wise storage is implemented. In the field of engineering and natural sciences, numerical methods with matrices have been highly developed over decades, and extensive software is available worldwide. Therefore, this convention of column-wise storage, which is common in theory and practice, will be continued here. The following sections describe corresponding software classes for processing matrices and vectors based on the object-oriented paradigm, which is also the foundation of the C++ programming language [Breymann 1997].

U. F. Meissner, *Tensor Calculus with Object-Oriented Matrices for Numerical Methods in Mechanics and Engineering*, https://doi.org/10.1007/978-3-031-59302-4_3

Object-oriented programming provides excellent prerequisites for modeling processes, physical objects, and software design, allowing for the analysis of complex structures and their effective representation in well-organized models and software modules [Hartmann (Hrsg.) 2000], [Rumbaugh/Blaha/Premerlani/Eddy/Lorensen 1991]. Objects, whose properties and methods are specified in class structures, play a central role in this context. Encapsulation of their properties in privately declared sections offers runtime-generated objects the necessary protection against arbitrary destruction or unintentional modification. It is also very advantageous that the properties of classes, consisting of member variables and functions, can be inherited within the hierarchical class structure, making them shareable without the need for repeated definitions. Another fundamental principle being advantageously used here concerns communication between objects at run time. Objects communicate with each other solely through well-defined interfaces. In C++, these are the public member functions (methods) bound to the respective class object (instance) when called using the dot '.' or arrow '->' access operators.

This proven structure of encapsulation, inheritance, and communication is inherently modeled after the behavior and structure of biological individuals and their organization in societies [Coad/Yourdon 1990]. Here, it is applied to matrix objects using the object-oriented programming language C++, detailing the construction and modeling of matrices with the explanation of the classes.

The C++ programming language provides ideal conditions for implementation. During run time, multidimensional arrays for matrix storage can be dynamically created and deleted by temporarily providing the corresponding main memory from the operating system. Moreover, the usual operators for arithmetic and management of scalar quantities and arrays can be overloaded, adapted to the new matrix classes, and made available to programming as an extension of the language syntax. The latter makes working with matrices and vectors intuitive and compact, and it also opens up the possibility to translate index notation into corresponding algorithms for multidimensional matrix operations. This eliminates many restrictive constraints imposed by two-dimensional symbolic matrix calculations.

3.1 Indexing and Storage of Matrices

As an introduction, this section addresses the handling of multidimensional matrices and their storage techniques. Because conventional matrix representation is traditionally based on two-dimensional notation, dealing with multidimensional matrices may be unfamiliar. On the other hand, programming languages designed for mathematical and scientific tasks provide multidimensional arrays to store and process also hypermatrices in addition to vectors and rectangular matrices. In the context of object-oriented matrix processing, some agreements and rules are established here to subsequently implement them into appropriate objects and object-oriented methods for matrix operations.

The storage of one- and two-dimensional matrices was already discussed in Section 2.1. Accordingly, it is assumed here that all matrices are stored column-wise; i.e., for a two-dimensional matrix with dimensions $(N_i \times N_k)$:

$$A_{ik} = \begin{pmatrix} a_{00} & a_{01} \\ a_{10} & a_{11} \\ a_{20} & a_{21} \end{pmatrix}$$

the first index runs through rows $i = 0, ..., N_{i-1}$, and the second index runs through columns $k = 0, ..., N_{k-1}$. Consequently, the elements of this matrix can also be stored in a one-dimensional array with dimensions $(N_L = N_i * N_k)$

$$B_L := \left[a_{00}\ a_{10}\ a_{20}\ a_{01}\ a_{11}\ a_{21} \right],$$

where index L sequentially runs through all elements $L = 0, ..., N_{L-1}$ to identify individual elements.
This well-known principle of sequential allocation is applied to multidimensional matrices below. In conventional von Neumann computers, whose physical main memory is linear and sequentially addressable, data mapping and storage also follow such a principle.

Here, application and explanation are limited to rectangular matrices. For differently structured matrices, such as triangular matrices, band matrices, skyline matrices, or other special forms, very specific storage and access methods, including considerations of size and efficiency, apply. These methods may be extracted from the literature and integrated into the object-oriented concept.

Rules for Bundling of Indices

For rectangular matrices, the fundamental principle here is that all matrices are stored column-wise. The described assignment process is allowed even if the matrices have the same name and remain in the same memory location. Only the dimensions of the matrix are changed. This change in indexing is referred to as ***bundling up adjacent indices***, and it is indicated as follows:

$$\boxed{\begin{array}{c} A_L := A \underbrace{ik}_{L} \quad or \quad C_{iM} := C_i \underbrace{klm}_{M} \\ with\ the\ new\ dimensions \\ N_L = N_i * N_k \ resp.\ N_M = N_k * N_l * N_m \end{array}} \,. \tag{3.1.1}$$

This specific ***assignment***, causes no change in content or memory relocation of matrix elements. From an organizational perspective, only the dimensions of one and the same matrix are modified. For use in algorithms, it is envisioned that such changes can be either permanent or temporary. The current state will therefore be recorded within the protected private data area of the matrix object to monitor unauthorized memory access during computation and memory operations.

To further illustrate this rule, consider a ***four-dimensional hypermatrix*** $\boldsymbol{K_{iklm}}$ with dimensions $(N_i \times N_k \times N_l \times N_m)$, as shown in ***Section 3.2, Table 3.1***. To store such a matrix as a two-dimensional matrix with N_Z rows and N_S columns, the indices can be bundled up as follows:

$$K_{ZS} := K \underbrace{ik}_{Z}\underbrace{lm}_{S}$$

with the new dimensions $N_Z = N_i \times N_k$ and $N_S = N_l \times N_m$. This results in the column-wise representation of the two-dimensional $[N_z \times N_s]$-matrix

$$\boxed{\mathrm{K}_{ZS} := \left[K_{Z0}\ K_{Z1}\ K_{Z2}\ K_{Z3}\ K_{Z4}\ K_{Z5}\ \ldots .\ K_{ZN_s-1} \right]} \,. \tag{3.1.2}$$

If the indices Z and S are further bundled up

$$K_\Omega := K \underbrace{ZS}_{\Omega} ,$$

the sequential one-dimensional storage scheme follows with dimension $N_\Omega = N_Z * N_S$

$$\begin{aligned} K_\Omega := \big[& K_{0000}\ K_{1000}\ K_{0100}\ K_{1100}\ K_{0200}\ K_{1200}\ K_{0300}\ K_{1300} \mid \\ & K_{0010}\ K_{1010}\ K_{0110}\ K_{1110}\ K_{0210}\ K_{1210}\ K_{0310}\ K_{1310} \mid \\ & K_{0001}\ \ldots\ldots\ldots \big] . \end{aligned}$$

These mappings are bijective, as there are unique rules for converting the indices of the ***four-dimensional hypermatrix K_{iklm} into the rectangular matrix K_{ZS}*** and vice versa:

$$\boxed{\begin{array}{l} \textit{for the } \textbf{mapping } K_{ZS} := K_{\underbrace{ik}_{Z}\underbrace{lm}_{S}} \\ \textit{the assignments} \\ Z = k * N_i + i \;\; and \;\; S = m * N_l + l\,, \\ \\ \textit{and for its } \textbf{reverse mapping } K_{\underbrace{ik}_{Z}\underbrace{lm}_{S}} := K_{ZS} \\ \textit{the assignments} \\ k = Entier(Z/N_i) \;\; and \;\; i = Z - k * N_i \\ m = Entier(S/N_l) \;\; and \;\; l = S - m * N_l \end{array}} \; . \tag{3.1.3}$$

Similarly, for the conversion of the ***two-dimensional matrix K_{ZS} into the vector K_Ω*** and vice versa, the following rules apply:

$$\boxed{\begin{array}{l} \textit{for the } \textbf{mappings } K_\Omega := K_{\underbrace{ZS}_{\Omega}} \;\; and \;\; K_{\underbrace{ZS}_{\Omega}} := K_\Omega \\ \textit{the assignments} \\ \Omega = S * N_Z + Z \\ S = Entier(\Omega/N_Z) \;\; and \;\; Z = \Omega - S * N_Z \end{array}} \; . \tag{3.1.4}$$

When applied in numerical programs, organizing to bundle up indices in this way brings significant advantages in numerical computations with multidimensional matrices. Complex operations can be efficiently reduced to basic matrix operations for which well-tested program routines are available. This approach is further explained in the following examples, with the inclusion of zero indices as mentioned already in Section (2.1.2).

Example ***Addition/Subtraction***: As basic operation the algorithm

$$C_i := A_i \pm B_i$$

should be available. This operation can be applied by bundling up indices of multidimensional matrices as follows:

$$c_{\underbrace{ikl}_{I}} := a_{\underbrace{ikl}_{I}} \pm b_{\underbrace{ikl}_{I}} \; . \tag{3.1.5}$$

Example ***Multiplication***: As basic operation the algorithm

$$C_{ikmn} := A_{i\mathbf{l}k} * B_{m\mathbf{l}n}$$

should be available. This operation can be used by bundling up indices and inserting zero indices for the multiplication of multidimensional matrices

$$F_{ik} = K_{iklm} * V_{lm}$$

which can be utilized as follows:

$$\mathrm{F}\underbrace{{}_{ik}}_{Z}{}_{000} := K\underbrace{{}_{ik}}_{Z}\underbrace{{}_{lm}}_{\boldsymbol{S}}{}_{0} * V_{0}\underbrace{{}_{lm}}_{\boldsymbol{S}}{}_{0}\,. \tag{3.1.6}$$

Example ***Determinant***: The basic operation involves again the multiplication of three-dimensional matrices

$$C_{ikmn} := A_{\boldsymbol{il}k} * B_{m\boldsymbol{l}n}\,.$$

For the determinant calculation following (2.2.9), with three vectors

$$A0_i := A_{i0} \; and \; A1_i := A_{i1} \; and \; A2_i := A_{i2}$$

the multiplication can be performed with the three-dimensional e-matrix

$$det(A) = e_{\boldsymbol{ikl}} * A0_{\boldsymbol{i}} * A1_{\boldsymbol{k}} * A2_{\boldsymbol{l}}$$

in multiple steps:

$$\begin{aligned} h_{il00} &:= e_{i\boldsymbol{k}l} * A1_{0\boldsymbol{k}0}\,, \\ k_{i000} &:= h_{i\boldsymbol{l}}\underbrace{{}_{00}}_{0} * A2_{0\boldsymbol{l}0}\,, \\ det(a) = h_{0000} &:= k_{0\boldsymbol{i}}\underbrace{{}_{000}}_{0} * A_{0\boldsymbol{i}0}\,. \end{aligned} \tag{3.1.7}$$

Of cource, the numerical efficiency of such formulations needs to be assessed and optimized in each specific case when implementing them into optimal algorithms.

3.2 Four-dimensional Hypermatrix

After discussing the storage and handling of multidimensional matrices in Section 3.1, we will now present such a four-dimensional $(2\times4\times2\times4)$-hypermatrix K_{iklm} in a tabular form with its indices following the structure defined in Equation (3.1.6). We also explicitly outline the mapping rules for consolidating the indices to convert them into the two-dimensional matrix K_{ZS} or the one-dimensional vector K_{Ω}.

The principle of column-wise storage in the two-dimensional storage format K_{ZS} is illustrated based on the indices listed on the left and top edges of Table 3.1:
- First, index i iterates through the matrix elements, followed by the second index k, then the third index l, and finally the last index m.
This sequence also determines the sequential assignment of all matrix elements to the one-dimensional storage format K_{Ω} at the bottom of the table.

Access to the identical matrix elements in the different storage formats K_{iklm}, K_{ZK}, and K_{Ω} is done by specifying the indices $iklm$, ZK, or Ω. This approach also allows for extracting or inserting individual parts into the hypermatrix. In Section 3.3.1, specific methods (submatrix_out, submatrix_in, or partialmatrix_out, partialmatrix_in are declared to access such matrix parts in object-oriented algorithms accordingly.
As an example, a table is presented on the following page containing sixteen such (4×4) submatrices, which were addressed using row/column indices k and m to illustrate their location in the overall matrix.

In the application of the FE-method, the processes of storing and retrieving data are of central importance because large matrices at the global system level often need to be assembled or disassembled concerning the smaller matrices at the individual element level. This table example can help in understanding these operations and accesses. It primarily demonstrates the technique of consolidating multiple indices for the application of matrix methods.

Table 3.1: Structure of $(2\times4\times2\times4)$ -Hypermatrix K_{iklm}

			l	0	1	0	1	0	1	0	1
			m	0		1		2		3	
	i	k	Z | S	0	1	2	3	4	5	6	7
	0	0	0	K_{0000}	K_{0010}	K_{0001}	K_{0011}	K_{0002}	K_{0012}	K_{0003}	K_{0013}
	1		1	K_{1000}	K_{1010}	K_{1001}	K_{1011}	K_{1002}	K_{1012}	K_{1003}	K_{1013}
	0	1	2	K_{0100}	K_{0110}	K_{0101}	K_{0111}	K_{0102}	K_{0112}	K_{0103}	K_{0113}
$K_{ZS} :=\ K_{iklm} :=$	1		3	K_{1100}	K_{1110}	K_{1101}	K_{1111}	K_{1102}	K_{1112}	K_{1103}	K_{1113}
	0	2	4	K_{0200}	K_{0210}	K_{0201}	K_{0211}	K_{0202}	K_{0212}	K_{0203}	K_{0213}
	1		5	K_{1200}	K_{1210}	K_{1201}	K_{1211}	K_{1202}	K_{1212}	K_{1203}	K_{1213}
	0	3	6	K_{0300}	K_{0310}	K_{0301}	K_{0311}	K_{0302}	K_{0312}	K_{0303}	K_{0313}
	1		7	K_{1300}	K_{1310}	K_{1301}	K_{1311}	K_{1302}	K_{1312}	K_{1303}	K_{1313}

Z= $k*N_i + i$
S= $m*N_l + l$

k=Entier (Z/N_i) m=Entier (S/N_l)
i = $Z-k*N_i$ l = $S-m*N_l$

	Z	0	1	2	3	4	5	6	7	0	1	2	3	..
	S	0								1				
	Ω	0	1	2	3	4	5	6	7	8	9	10	11	..
$K_{ZS} :=$	$K_\Omega :=$	K_{0000}	K_{1000}	K_{0100}	K_{1100}	K_{0200}	K_{1200}	K_{0300}	K_{1300}	K_{0010}	K_{1010}	K_{0110}	K_{1110}	..

$$\Omega = S*N_z + Z$$

S=Entier (Ω/N_z)
Z= $\Omega - S*N_z$

Submatrices of $(2\times4\times2\times4)$*-Hypermatrix* K_{iklm}

$$\begin{aligned}
&\boldsymbol{K}_{i0l0} := \begin{pmatrix} K_{0000} & K_{0010} \\ K_{1000} & K_{1010} \end{pmatrix} \quad \boldsymbol{K}_{i0l1} := \begin{pmatrix} K_{0001} & K_{0011} \\ K_{1001} & K_{1011} \end{pmatrix} \\
&\boldsymbol{K}_{i1l0} := \begin{pmatrix} K_{0100} & K_{0110} \\ K_{1100} & K_{1110} \end{pmatrix} \quad \boldsymbol{K}_{i1l1} := \begin{pmatrix} K_{0101} & K_{0111} \\ K_{1101} & K_{1111} \end{pmatrix} \\
&\boldsymbol{K}_{i2l0} := \begin{pmatrix} K_{0200} & K_{0210} \\ K_{1200} & K_{1210} \end{pmatrix} \quad \boldsymbol{K}_{i2l1} := \begin{pmatrix} K_{0201} & K_{0211} \\ K_{1201} & K_{1211} \end{pmatrix} \\
&\boldsymbol{K}_{i3l0} := \begin{pmatrix} K_{0300} & K_{0310} \\ K_{1300} & K_{1310} \end{pmatrix} \quad \boldsymbol{K}_{i3l1} := \begin{pmatrix} K_{0301} & K_{0311} \\ K_{1301} & K_{1311} \end{pmatrix} \\
&\boldsymbol{K}_{i0l2} := \begin{pmatrix} K_{0002} & K_{0012} \\ K_{1002} & K_{1012} \end{pmatrix} \quad \boldsymbol{K}_{i0l3} := \begin{pmatrix} K_{0003} & K_{0013} \\ K_{1003} & K_{1013} \end{pmatrix} \\
&\boldsymbol{K}_{i1l2} := \begin{pmatrix} K_{0102} & K_{0112} \\ K_{1102} & K_{1112} \end{pmatrix} \quad \boldsymbol{K}_{i1l3} := \begin{pmatrix} K_{0103} & K_{0113} \\ K_{1103} & K_{1113} \end{pmatrix} \\
&\boldsymbol{K}_{i2l2} := \begin{pmatrix} K_{0202} & K_{0212} \\ K_{1202} & K_{1212} \end{pmatrix} \quad \boldsymbol{K}_{i2l3} := \begin{pmatrix} K_{0203} & K_{0213} \\ K_{1203} & K_{1213} \end{pmatrix} \\
&\boldsymbol{K}_{i3l2} := \begin{pmatrix} K_{0302} & K_{0312} \\ K_{1302} & K_{1312} \end{pmatrix} \quad \boldsymbol{K}_{i3l3} := \begin{pmatrix} K_{0303} & K_{0313} \\ K_{1303} & K_{1313} \end{pmatrix}
\end{aligned} \tag{3.2.1}$$

3.3 Base Class MATRIZEN

The class MATRIZEN[1] [Meissner 2023] is an abstract base class that does not have constructors for creating class instances as matrix objects. Its primary purpose is to declare common properties that should apply to possible derived classes (e.g. MATRIX and VEKTOR[2]) and be inherited. In this case, these are the specified member functions that can be accessed as public, and their documentation is provided below. Additionally, derived classes also have access to hidden data components for shared use. In the protected section, some member variables are already declared, which are necessary for the setup and common management of matrix instances. These are hidden from external access through encapsulation and are not accessible for users. However, at the end of the class documentation, they are exceptionally disclosed here for a better understanding of the class structure.

[1] engl. MATRICES
[2] engl. VECTOR

3.3.1 General Matrix Methods

In this design, within the protected data area, a pointer 'double *matrix' is provided, which points to the domain of a one-dimensional array with a total dimension of dimmax. This is where the four-dimensional MATRIX object with dimensions dimi, dimj, dimk, diml is stored once it has been created by the constructor of a derived class. The data member "type" allows the management of various states of the object. Temporary matrix objects are used solely as storage for intermediate calculations or states, and should be deleted after the corresponding operation. For temporary incrementation of the pointer *matrix to the MATRIX object, which must be preserved, an additional pointer "*current_element" is available for shared internal access. Furthermore, within the local access area of derived classes additional properties may be specified for specific purposes.

Some arguments of member functions have default values, allowing them to be called with a shortened list of data items for user convenience. Some functions, marked as virtual, such as the destructor '~' for deleting dynamic matrices, do not have a particular implementation at this level because their definition is left to derived classes. All declared functions cannot be used before instances (objects) have been defined in a certain class and are created in a computational program, to which they are bound when called using the dot operator '.' or the arrow operator '->'. The base class merely declares interfaces for potential objects. In this case, they apply to accessing matrix indices, individual matrix elements or parts of matrices, to input/output operations with cin '»' and cout '«', the C++ assignment operator '=' (equivalent to the mathematical assignment symbol ':=') as well as elementary arithmetic operators, and other management operations such as determining or changing matrix dimensions using the specific member function 'dimension'.

The base class MATRIZEN more over has some protected member functions, but they are not of relevance for understanding here, cf. [Meissner 2023]. The declared member functions may be modified and further specialized in the later on assigned derived classes as needed.

Declaration file (header file): Matrizen.h

Listing 3.1: Abstract Base Class MATRIZEN

```
// ========== MATRIZEN base class for general matrices ==============
// Declaration file (header file)
class MATRIZEN
// ======== Declaration of prototypes ===============================
{
// MATRIZEN mat(i,j,k,l) are stored column-wise !!!
// Indexing of i,j,k,l starts at 0 !!!
// U. Breymann: C++. Carl Hanser Verlag (1993), pp. 186 ff
// -----------------------------------------------------------------
// Copyright: Univ.-Prof. Dr.-Ing. Udo F. Meissner, Erkrath/Darmstadt
// Advanced teaching and study program - Commercial use not permitted.
// The author assumes no liability for incompleteness and errors!
// Version:  2023
```

```
// Compiler: C++
// ===================================================================
//                    Public member functions
//                    =======================
  public:
// =======
//
// Lock matrix to prevent deletion -> mat.lock()
// ==============================================
  void sperren();
  void lock   ();
//
// Unlock matrix to allow deletion -> mat.unlock()
// ================================================
  void entsperren();
  void unlock    ();
//
// Search for VALUE
// ================
// Arguments:
// knonr            : value searched for
// ni,nj,nk,nl      : length of matrix dimensions
//                  : -1 evaluate length automatically
// di,dj,dk,dl      : step size for value search
// Function result: 1 knonr was found
//                   -1 knonr was not found
int wert_suchen(double knonr, int& i, int& j, int& k, int& l,
    int ni = 1, int di = 1, int nj = 1, int dj = 1, int nk = 1,
    int dk = 1, int nl = 1, int dl = 1);
int value_search(double knonr, int& i, int& j, int& k, int& l,
    int ni = 1, int di = 1, int nj = 1, int dj = 1, int nk = 1,
    int dk = 1, int nl = 1, int dl = 1);
//
// ===================================================================
//                  Public, virtual member functions
//                  ================================
  public:
// =======
//
// Virtual destructor -> delete &mat
// ==================================
virtual ~MATRIZEN();
//
// Delete temporary matrix -> mat.delete_temporary_matrix()
// ==========================================================
virtual void temporaere_matrix_loeschen() = 0;
virtual void delete_temporary_matrix()   = 0;
//
// ===================================================================
//         Public member functions (with temporary matrices)
//         ================================================
  public:
// =======
//
```

```
// Type conversion operator
// -> number = double(matrix) (assign the first matrix element)
// ============================================================
operator double ();                        // Overloaded operator double
//
// Matrix index -> (i)
// ====================
double &operator () (int i);                  // Overloaded operator ()
// Matrix index -> (i,j)
// ====================
double &operator () (int i, int j);           // Overloaded operator ()
// Matrix index -> (i,j,k)
// ======================
double &operator () (int i, int j, int k); // Overloaded operator ()
// Matrix index -> (i,j,k,l)
// ========================
double &operator () (int i, int j, int k, int l);
                                              // Overloaded operator ()
//
// Matrix element
// -> mat.matrix_element(i,j,k,l) compare with mat(i,j,k,l)
// ========================================================
double &element(int i=0, int j=0, int k=0, int l=0);
//
// Get the current matrix element and increment the element pointer
// -> mat++
// ================================================================
// by explicitly calling
// -> mat.operator ++ (distance)
// the element pointer can be incremented by several elements,
// the element pointer is only changed by the constructor, destructor,
// and the operators '()', '--' !
// ----------------------------------------------------------------
double &operator ++ (int distance);   // Overloaded operator ++
//
// Get current matrix element and decrement the element pointer
// -> mat--
// ============================================================
// By explicitly calling
// -> mat.operator -- (distance)
// the element pointer can be decremented by several elements,
// The element pointer is only changed by the constructor, destructor,
// and operators '()', operator '++' !
// ------------------------------------------------------------
double &operator -- (int distance);   // Overloaded operator --
//
// Temporarily change matrix dimensions
// -> mat.dimension(i,j,k,l)
// ================================
// the bundeling of indices (redimensioning) is actually needed
// for the matrix multiplication mat1*mat2,
// therefore, the redimensioning only applies
//                                    until the next matrix operation!
// afterwards, the original dimensions are restored !!!
```

```
// ------------------------------------------------------------------
MATRIZEN &dimension(int ni=1, int nj=1, int nk=1, int nl=1);
//
// Determine matrix dimensions
// -> mat.determine_dimension(ni,nj,nk,nl)
// ========================================
// Result: dim=number of matrix elements (allocated memory)
   int dimension_ermitteln(int &ni, int &nj, int &nk, int &nl);
   int determine_dimension(int &ni, int &nj, int &nk, int &nl);
//
// Permanently change matrix dimensions
// -> mat.change_dimension(i,j,k,l)
// ======================================
   MATRIZEN &dimension_aendern(int ni=1, int nj=1, int nk=1,int nl=1);
   MATRIZEN &change_dimension (int ni=1, int nj=1, int nk=1,int nl=1);
//
// Get number of rows/columns
// -> mat.rows_columns(ni,nj,nk,nl,dim)
// ======================================
// Result: Number of rows/columns=sqrt(ni*nj*nk*nl)
//         for square matrices,
//         otherwise -1
   int zeilen_spalten(int &ni, int &nj, int &nk, int &nl);
   int rows_columns  (int &ni, int &nj, int &nk, int &nl);
//
// Extract partial matrix
// -> ...=mat.partialmatrix_out(i,j, ni,nj, k,l, nk,nl)
// ====================================================
// partialmatrix_out retrieves consecutive matrix elements starting at
// index mat(i,j,k,l), which are stored
//                                  in the partial matrix(ni,nj,nk,nl)
// ------------------------------------------------------------------
   MATRIZEN &teilmatrix_heraus(int i=0, int j=0, int ni=1, int nj=1,
                               int k=0, int l=0, int nk=1, int nl=1);
   MATRIZEN &partialmatrix_out(int i=0, int j=0, int ni=1, int nj=1,
                               int k=0, int l=0, int nk=1, int nl=1);
//
// Insert partial matrix
// -> mat.partialmatrix_in(i,j,k,l)=...
// ======================================
// partialmatrix_in allows storing the partial matrix starting
//                                           at index mat(i,j,k,l)
// ------------------------------------------------------------------
   MATRIZEN &teilmatrix_hinein(int i=0, int j=0, int k=0, int l=0);
   MATRIZEN &partialmatrix_in (int i=0, int j=0, int k=0, int l=0);
//
// Extract submatrix
// -> ...= mat.submatrix_out(i,ni, j,nj, k,nk, l,nl)
// ================================================
// submatrix_out retrieves the submatrix starting at index mat(i,j,k,l
   )
//
   ------------------------------------------------------------------
```

```
   MATRIZEN &untermatrix_heraus(int i=0, int ni=1, int j=0, int nj=1,
                                int k=0, int nk=1, int l=0, int nl=1);
   MATRIZEN &submatrix_out     (int i=0, int ni=1, int j=0, int nj=1,
                                int k=0, int nk=1, int l=0, int nl=1);
//
// Insert submatrix
// -> mat.submatrix_in(i,j,k,l)=...
// ================================
// submatrix_in allows storing the submatrix starting
//                                                at index mat(i,j,k,l)
// ----------------------------------------------------------------------
   MATRIZEN &untermatrix_hinein(int i=0, int j=0, int k=0, int l=0);
   MATRIZEN &submatrix_in      (int i=0, int j=0, int k=0, int l=0);
//
// Matrix output -> cout<<mat
// ==========================
friend std::ostream& operator << (std::ostream& s, MATRIZEN& mat);
//                                              // Overloaded operator <<
//
// Matrix input -> cin>>mat
// ========================
friend std::istream &operator >> (std::istream& s, MATRIZEN& mat);
//                                              // Overloaded operator >>
//
// Set all matrix elements -> mat = number
// ========================================
MATRIZEN &operator  = (double number);   // Overloaded operator =
// Set all matrix elements -> mat+= number
// ========================================
MATRIZEN &operator += (double number);   // Overloaded operator +=
// Set all matrix elements -> mat-= number
// ========================================
MATRIZEN &operator -= (double number);   // Overloaded operator -=
//
// Assign matrix -> mat1 = mat2 (copy matrix content only)
// ============================
MATRIZEN &operator  = (MATRIZEN &mat2);  // Overloaded operator  =
// Assign matrix -> mat1+= mat2 (copy matrix content only)
// ============================
MATRIZEN &operator += (MATRIZEN &mat2);  // Overloaded operator +=
// Assign matrix -> mat1-= mat2 (copy matrix content only)
// ============================
MATRIZEN &operator -= (MATRIZEN &mat2);  // Overloaded operator -=
//
// Positive sign -> +mat
// =====================
MATRIZEN &operator + ();                 // Overloaded operator +
//
// Addition with scalar -> mat+number
// ==================================
MATRIZEN &operator + (double number);    // Overloaded operator +
//
// Addition with scalar -> number+mat
// ==================================
```

```
friend MATRIZEN &operator + (double number, MATRIZEN &mat);
                                           // Overloaded operator +
//
// Addition -> mat1+mat2 or mat1+vec2 or vec1+mat2
// ===============================================
MATRIZEN &operator + (MATRIZEN &mat2);    // Overloaded operator +
//
// Negative sign -> -mat
// ====================
MATRIZEN &operator - ();                  // Overloaded operator -
//
// Subtraction with scalar -> mat-number
// ====================================
MATRIZEN &operator - (double number);     // Overloaded operator -
//
// Subtraction with scalar -> number-mat
// ====================================
friend MATRIZEN &operator - (double number, MATRIZEN &mat);
                                           // Overloaded operator -
//
// Subtraction -> mat1-mat2 or mat1-vec2 or vec1-mat2
// ==================================================
MATRIZEN &operator - (MATRIZEN &mat2);    // Overloaded operator -
//
// Multiplication with scalar -> mat*number
// =======================================
MATRIZEN &operator * (double number);      // Overloaded operator *
//
// Multiplication with scalar -> number*mat
// =======================================
friend MATRIZEN &operator * (double number, MATRIZEN &mat);
                                           // Overloaded operator *
//
// Matrix multiplication -> matc(i,j,k,l) = mat1(i,s,j) * mat2(k,s,l)
// => To ensure uniqueness, this operation must be regulary
//    executed within a single statement !
// =================================================================
MATRIZEN &operator * (MATRIZEN &mat2);    // Overloaded operator *
//
// 'Matrix division'      -> MATx       = MATr      / MATk
//                        -> matx(s,n) = matr(i,n) / matk(i,s)
// ==========================================================
// Linear equation solution -> matk(i,s) * matx(s,n) = matr(i,n)
// The staggered form of matk is stored internally,
//                        to process further right-hand sides matr
// =============================================================
MATRIZEN& operator / (MATRIZEN &MATk);   // Overloaded operator /
//
// Division with scalar -> mat/number
// =================================
MATRIZEN &operator / (double number);    // Overloaded operator /
//
// Absolute value of a matrix -> mat.absolute_value()
// ==================================================
```

```
   double betrag         ();
   double absolute_value();
//
// Normalize matrix -> mat.normalize()
// ==================================
   MATRIZEN &normieren();
   MATRIZEN &normalize();
//
// Normalize matrix -> mat.normalize(absolute_value)
// ================================================
   MATRIZEN &normieren(double &absolute_value);
   MATRIZEN &normalize(double &absolute_value);
//
// ========================================================
// ...  further member functions, cf. [Meissner 2023]  ...
// ========================================================
//
//        Protected data components -> Matrix descriptor
//        ==============================================
   protected:
// ==========
//
// Type of the matrix ->  2 = permanent, locked against deletion
// ------------------ ->  1 = permanent
//                    ->  0 = temporary with old memory
//                    -> -1 = temporary with new memory
//  type
int typ;
//
// Dimensions of the 4D-matrix mat(dimi,dimj,dimk,diml)
// ----------
int dimi, dimj, dimk, diml;
//
// Total dimension dimmax=dimi*dimj*dimk*diml
// ----------------                         of the vector matrix[dimmax]
int dimmax;
//
// Pointer on matrix[dimmax] for storing
//                                   the matrix mat(dimi,dimj,dimk,diml)
double *matrix;
//
// Pointer to the current matrix element
// for the increment operators &operator ++ (), &operator -- ()
//      *current_element
double *aktuelles_element;
};
// =================================================================
```

3.3.2 Member Functions for Linear Equation Systems

In the following, the base class MATRIZEN[1] is amended [Meissner 2023] by some public ***methods for solving linear equation systems***, as discussed in Section 2.3. These member functions are generally valid, but have been adapted for finite element method purposes, as described in Section 2.3.

First, the singular coefficient matrix Kmat, such as the positive-semidefinite stiffness matrix (2.3.7), is modified with the function call Kmat.***support***, incorporating the prescribed geometrical boundary conditions by 'striking out rows and columns' according to the support matrix Smat (2.3.9), in order to obtain the regular matrix version of Amat.

The triangular decomposition (2.3.15) of the reduced coefficient matrix Amat (2.3.11) is then performed for symmetric equation systems by calling the functions Amat.***cholesky_decomposition*** resp. Amat.***gauss_decomposition*** with the result of the upper matrix Umat. For unsymmetric equation systems, it is also done by calling the function Amat.***gauss_decomposition, resulting with two matrices***, the upper matrix Umat and the lower matrix Lmat.

The decomposition of the right-hand side Rmat of the reduced equation system (2.3.11) is carried out with the function call Rmat.***decomposition*** by forward substitution (2.3.15), according to the parameter symdef, either with Cholesky or Gauss elimination, using the lower triangular matrix Lmat.

Finally, the solution of the linear equation system is determined by backward substitution (2.3.15) with the function call Xmat.***solution***, using the upper triangular matrix Umat and the decomposed right-hand side Ymat, .

For the determination of the inverse matrix of the square matrix Amat, the function call Amat.***invert*** is available, optionally by Cholesky or Gauss decomposition, with the result of the inverted matrix Zmat .

The base class MATRIZEN has set a general framework for the declaration of those member functions that may be tailored further on to specific matrix forms, such as band matrices, skyline matrices, etc. through specialization in additional derived classes.

In this case of the rectangular matrices, the necessary specialization is complemented within the derived class MATRIX by the definition of private data and member functions that are protected from external access and therefore remain hidden.

Supplement to the declaration file (header file): Matrizen.h

Listing 3.2: Base Class MATRIZEN (Linear Equation Systems)

```
// ========== MATRIZEN base class for general matrices ===========
```

[1] engl. MATRICES

```
// Declaration file (header file)
// Public member functions for solving of linear equation systems
// ===============================================================
// Copyright: Univ.-Prof. Dr.-Ing. Udo F. Meissner, Erkrath/Darmstadt
// Advanced teaching and study program - Commercial use not permitted.
// The author assumes no liability for incompleteness and errors!
// Version:  2023
// Compiler: C++
// ===============================================================
//                Public member functions
//                =======================
   public:
// =======
//
// Symmetric coefficient matrix - Gauss decomposition
// -> Amat.gauss_decomposition(Umat)
// ==================================================
// Convert a symmetric coefficient matrix
// to the Gauss staggered form
// A(mk,mk) = Matrix with fully stored coefficient matrix
// U(mk,mk) = Upper matrix for the Gauss triangel
//            (U can be the same as A)
// mk       = Rows/Columns of A, U
// Result:    Determinant of matrix A
// ------------------------------------------------------------------
double nach_gauss_staffeln     (MATRIZEN& U);
double       gauss_decomposition(MATRIZEN& U);
//
// Symmetric coefficient matrix - Cholesky decomposition
// -> Amat.cholesky_decomposition(Umat)
// =====================================================
// Convert a symmetric and positive-definite coefficient matrix
// to the Cholesky staggered form
// A(mk,mk) = Matrix with fully stored coefficient matrix
// U(mk,mk) = Upper matrix of the Cholesky triangle
//            (U can be the same as A)
// mk       = Rows/Columns of A, U
// Result:    Determinant of matrix A
// ------------------------------------------------------------------
double nach_cholesky_staffeln     (MATRIZEN& U);
double       cholesky_decomposition(MATRIZEN& U);
//
// Unsymmetric coefficient matrix - Gauss decomposition
// -> Amat.gauss_decomposition(Umat,Lmat)
// =====================================================
// Convert an unsymmetric coefficient matrix
// to the Gauss factorized forms
// A(mk,mk) = Matrix with fully stored coefficient matrix
// U(mk,mk) = Upper matrix for the Gauss triangel
//            (U can be the same as A)
// L(mk,mk) = Lower matrix for the Gauss triangel
//            (L can be the same as U)
// mk       = Rows/Columns of A, U, L
// Result:    Determinant of matrix A
```

```
// -----------------------------------------------------------------
double nach_gauss_staffeln     (MATRIZEN& U, MATRIZEN& L);
double       gauss_decomposition(MATRIZEN& U, MATRIZEN& L);
//
// Right-hand sides - Cholesky or Gauss decomposition
// -> Rmat.decomposition(Lmat,Ymat,symdef)
// ====================================================
// Convert the right-hand sides
// to the Cholesky resp. Gauss factorized form
// R(mb,zr) = Matrix with fully stored right-hand sides
// L(mb,mb) = Matrix with staggered coefficient matrix
//            (lower triangular matrix, if Gauss unsymmetric)
// Y(mb,zr) = Matrix for the factorized right-hand sides
//            (Y can be the same as R)
// mb       = Number of rows in L, R, Y
// zr       = Number of columns in right-hand sides R, Y
// symdef= 1: Cholesky decomposition
//       = 0: symmetrical Gauss decomposition
//       =-1: unsymmetricall Gauss decomposition
// -----------------------------------------------------------------
void staffeln      (MATRIZEN& L, MATRIZEN& Y, int symdef = -1);
void decomposition(MATRIZEN& L, MATRIZEN& Y, int symdef = -1);
//
// Solutions - Cholesky or Gauss backward substitution
// -> Xmat.solutions(Umat,Ymat,symdef)
// ====================================================
// Determine solutions for an equation system
// after Cholesky (symmetric) or after Gauss (symmetric) or
// after Gauss unsymmetric) decomposition from the staggered
// coefficient matrices and the factorized right-hand sides
// X(mb,zr) = Matrix for fully stored solutions
// U(mb,mb) = Matrix with staggered coefficient matrix
//            (upper triangular matrix, if Gauss unsymmetric)
// Y(mb,zr) = Matrix with factorized right-hand sides
//            (Y can be the same as X)
// mb       = Number of rows in U, X, Y
// zr       = Number of columns in X, Y
// symdef= 1: Cholesky decomposition
//       = 0: symmetrical Gauss decomposition
//       =-1: unsymmetrical Gauss decomposition
// -----------------------------------------------------------------
MATRIZEN& loesungen(MATRIZEN& U, MATRIZEN& Y, int symdef = -1);
MATRIZEN& solutions(MATRIZEN& U, MATRIZEN& Y, int symdef = -1);
//
// Inverse matrix - Cholesky or Gauss decomposition
//  of a symmetric or unsymmetric matrix
// -> Amat.invert(Zmat,symdef)
// ====================================================
// Invert the coefficient matrix by decomposition
// after Cholesky (only positive definite matrix) or Gauss
// A(mk,mk) = Matrix with fully stored coefficient matrix
// Z(mk,mk) = Matrix for the inverse
//            (Z can be the same as A)
// mk       = Rows/Columns of A, Z
```

```
// symdef= 1: Cholesky
//        = 0: Gauss symmetric
//        =-1: Gauss unsymmetric
// Result:    Determinant of matrix A
// ------------------------------------------------------------------
double invertieren(MATRIZEN &Z, int symdef = -1);
double invert     (MATRIZEN& Z, int symdef = -1);
//
// Incorporate boundary conditions into the coefficient matrix
// -> Kmat.support(Smat,Amat)
// ============================================================
// Incorporate boundary resp. support conditions
// K(mk,mk) = Matrix with fully stored coefficient matrix
// S(mk)    = Vector with support information
//    -1 = Rigid support (restricted degrees of freedom)
//     0 = No support  (unrestricted degrees of freedom)
//    >0 = Add spring constant to the main diagonal
// A(mk,mk) = Matrix for the supported coefficient matrix
//            (A can be the same as K)
// mk       = Rows/Columns of K, A
//            Rows of S
// ------------------------------------------------------------------
void lagern (MATRIZEN& S, MATRIZEN& A);
void support(MATRIZEN& S, MATRIZEN& A);
// ==================================================================
```

3.4 Matrix Classes MATRIX and intMATRIX

The class MATRIX [Meissner 2023] inherits all properties of the base class MATRIZEN through the declaration 'class MATRIX : public MATRIZEN'.

By use of the two constructors of the class MATRIX, one- to four-dimensional matrices with elements of type 'double' can be created. The matrix elements may be preloaded using an array in notation of the programming language C. For performing the general matrix multiplication (2.1.10), it is essential to use the member function 'dimension' which at runtime temporarily modifies the defined matrix dimensions to bundle up neighboring indices for the required order of the multiplicands. Input and output operations are performed using overloaded operators '>>' and '<<'.

Example:

```
double F8[8] = {0,1,2,3,4,5,6,7};  // preloaded C array F8
MATRIX M1(F8, 4, 2);            // static matrix M1, preloaded with F8

MATRIX& M2 = *new MATRIX(4, 2);     // dynamic matrix M2
        M2 = M2 - M1 * 5.0;         // assignment M2 := M2 - M1 * 5.0
cout << M1.dimension(4 * 2);        // temporarily bundled indices
cout << M1 << M2;                   // print
delete& M2;                         // delete using destructor '~'
```

Declaration file (header file): Matrix.h

Listing 3.3: Derived Class MATRIX

```
// ====== MATRIX class for rectangular matrices of type double ======
// Declaration file (header file)
class MATRIX : public MATRIZEN
// ====== Declaration of prototypes ==================================
{
// 4D-matrices mat(i,j,k,l) are stored column-wise !!!
// Indexing of i,j,k,l starts at 0 !!!
// See U. Breymann: C++. Carl Hanser Verlag (1993), p. 186 ff
// ---------------------------------------------------------------
// Copyright: Univ.-Prof. Dr.-Ing. Udo F. Meissner, Erkrath/Darmstadt
// Advanced teaching and study program - Commercial use not permitted.
// The author assumes no liability for incompleteness and errors!
// Version:  2023
// Compiler: C++
// ==================================================================
//Public member functions
//=====================
   public:
// =======
// Constructor -> MATRIX mat(ni,nj,nk,nl)
// ==================================
// ni, nj, nk, nl: Dimensions of the 4D matrix
// ---------------------------------------
MATRIX(int ni = 1, int nj = 1, int nk = 1, int nl = 1);
```

```
//
// Constructor -> MATRIX mat(C_array,ni,nj,nk,nl)
// ================================================
// C_array pre-initialized using an initialization list
// -> C_array [] = { ... }
// ni, nj, nk, nl: Dimensions of the 4D matrix
// ---------------------------------------------------------------
MATRIX(double* C_array = NULL, int ni = 1, int nj = 1, int nk = 1,
 int nl = 1);
//
// Copy constructor (copy the entire matrix)
// -> MATRIX mat1(&mat2)
// =========================================
MATRIX(MATRIX& mat2);
//
// Destructor -> delete &mat
// =========================
~MATRIX();
//
// Temporarily change matrix dimensions
// -> mat.dimension(i,j,k,l)
// =====================================
// included again here because MATRIX::transpose, MATRIX::trace,
// MATRIX::contraction do not belong to the base class MATRIZEN !!
// ---------------------------------------------------------------
// The bundeling of indices (redimensioning) is
// actually only required for matrix multiplication mat1*mat2.
// Therefore, the redimensioning applies only until the next
// matrix operation!
// After that, the original dimensions are restored!!!
// ---------------------------------------------------------------
MATRIX& dimension(int ni = 1, int nj = 1, int nk = 1, int nl = 1);
//
// Assign a value to all matrix elements -> mat = number
// =====================================================
// included again here because the assignment operator =
// cannot be inherited !!!
//  -------------------------------------------------------------
MATRIX& operator  = (double number);  // Overloaded operator =
//
// Assign matrix -> mat1 = mat2 (only copy matrix content)
// =======================================================
// included again here because the assignment operator =
// cannot be inherited !!!
// ------------------------------------------------------------
MATRIX& operator  = (MATRIZEN& mat2);  // Overloaded operator =
//
// Transpose matrix
// -> mat.transpose(index_0,index_1,index_2,index_3)
// =================================================
// Transpose stores all matrix elements around
// index_0, index_1, index_2, index_3 indicate
//which indices are swapped:
// e.g.,mat.transpose(3,2,1,0)
```

```
// -> mat1(l3,k2,j1,i0) <= mat(i0,j1,k2,l3)
// default case: mat.transpose()
// -> mat1(k,l,i,j) <= mat(i,j,k,l)
// 2D-matrices:  mat.transpose()
// -> possible without input parameters
// -----------------------------------------------------
MATRIZEN& transponieren(int index_0 = 2, int index_1 = 3,
int index_2 = 0, int index_3 = 1);
MATRIZEN& transpose    (int index_0 = 2, int index_1 = 3,
int index_2 = 0, int index_3 = 1);
//
// Contracting of a matrix
// -> mat.contract(index_0,index_1,index_2,index_3)
// =================================================
// index_0, index_1, index_2, index_3 indicate
// which indices are summed and which indices are swapped
// e.g.,mat.contract(0,0,3,2)
//-> mat1(l3,k2) <= mat(i0,i0,k2,l3)
// default case:
//mat.contract() -> mat1(i,k)<= mat(i,j,k,j)
// 2D-matrices:
//mat.contract() -> possible without input parameters
// -------------------------------------------------
MATRIZEN& verjuengen(int index_0 = 0, int index_1 = 1,
                     int index_2 = 2, int index_3 = 1);
MATRIZEN& contract  (int index_0 = 0, int index_1 = 1,
                     int index_2 = 2, int index_3 = 1);
//
// Trace of a matrix -> mat.trace(index_0,index_1,index_2,index_3)
// ==============================================================
// index_0, index_1, index_2, index_3 indicate
//which indices are summed
// e.g., mat.trace(0,0,2,2) -> generates trace <= mat(i0,i0,j2,j2)
// default case:
//  mat.trace() -> generates trace <= mat(i,j,i,j)
// 2D-matrices: mat.trace() -> possible without input parameters
// ------------------------------------------------------------
double spur (int index_0 = 0, int index_1 = 1,
             int index_2 = 0, int index_3 = 1);
double trace(int index_0 = 0, int index_1 = 1,
             int index_2 = 0, int index_3 = 1);
//
// Determinant of (3 x 3)- / (2 x 2)- / (1 x 1)-matrices
// -> Amat.determinant(n)
// ====================================================
// n = 3 / 2 / 1 dimension of matrix
// Result: determinant of matrix Amat
// ----------------------------------------------------
double determinante(int n = 1);
double determinant (int n = 1);
};
// ==============================================================
```

The class intMATRIX [Meissner 2023] is an independent base class that closely resembles the class MATRIX : public MATRIZEN in its properties. However, by use of templates with object type placeholders [Breymann 1993], it would be possible to combine these two classes, but this has been omitted here for the sake of clarity.

The two constructors of the intMATRIX class can be used to create one- to four-dimensional matrices with elements of type 'int'. The matrix elements may be preloaded using an array in notation of the programming language C. Input and output operations are performed using the overloaded operators '>>' and '<<'.

Example:

```
int I6[6] = {1,2,3,4,5,6};          // preloaded C array I6
intMATRIX N1(I6, 2, 3);         // static matrix N1, preloaded with I6

intMATRIX& N2 = *new intMATRIX(2, 3);  // dynamic matrix N2
           N2 = N2 + N1 * 7;        // assignment N2 := N2 + N1 * 7
cout << N1 << N2;                   // print
delete& N2;                         // delete using destructor '~'
```

Declaration file (header file): intMatrix.h

Listing 3.4: Base Class intMATRIX

```
// ===  intMATRIX class for rectangular matrices of type integer ===
// Declaration file (header file)
class intMATRIX
// ============== Declaration of prototypes ======================
{
// 4D-matrices mat(i,j,k,l) are stored column-wise !!!
// Indexing of i,j,k,l starts at 0 !!!
// See U. Breymann: C++. Carl Hanser Verlag (1993), p.186 ff
// ------------------------------------------------------------------
// Copyright: Univ.-Prof. Dr.-Ing. Udo F. Meissner, Erkrath/Darmstadt
// Advanced teaching and study program - Commercial use not permitted.
// The author assumes no liability for incompleteness and errors!
// Version:  2023
// Compiler: C++
// ==================================================================
//                    Public member functions
//                    ======================
   public:
// =======
//
// Constructor -> intMATRIX mat(ni,nj,nk,nl)
// =========================================
// ni, nj, nk, nl: Dimensions of the 4D matrix
// ------------------------------------------
intMATRIX (int ni=1, int nj=1, int nk=1, int nl=1);
//
// Constructor -> intMATRIX mat(C_array,ni,nj,nk,nl)
// =================================================
// C_array pre-initialized using an initialization list
```

```
// -> C_array [] = { ... }
// ni, nj, nk, nl: Dimensions of the 4D matrix
// -------------------------------------------------------------------
intMATRIX(int* C_array=NULL, int ni=1, int nj=1, int nk=1, int nl=1);
//
// Copy constructor (completely copy the entire matrix)
// -> intMATRIX  mat1(&mat2)
// ===================================================================
intMATRIX (intMATRIX  &mat2);
//
// Virtual destructor -> delete &mat
// ===================================
virtual ~intMATRIX ();
//
// Lock matrix to prevent deletion -> mat.lock()
// =============================================
   void sperren();
   void lock    ();
//
// Unlock matrix to allow deletion -> mat.unlock()
// ===============================================
   void entsperren();
   void unlock    ();
//
// Delete temporary matrix -> mat.delete_temporary_matrix()
// ========================================================
   void temporaere_matrix_loeschen();
   void delete_temporary_matrix   ();
//
// Search for a VALUE
// ==================
// Arguments:
// knonr: Value to search for
// ni,nj,nk,nl    : Dimensions of the matrix
//                : -1 automatically determine the length
// di,dj,dk,dl    : Step size for searching the matrix
// Function result:  1 knonr was found
//                  -1 knonr was not found
   int wert_suchen (int knonr, int &i, int &j, int &k, int &l,
       int ni=1, int di=1, int nj=1, int dj=1, int nk=1, int dk=1,
       int nl=1, int dl=1);
   int search_value(int knonr, int &i, int &j, int &k, int &l,
       int ni=1, int di=1, int nj=1, int dj=1, int nk=1, int dk=1,
       int nl=1, int dl=1);
// ===================================================================
//       Public member functions (with temporary matrices)
//       =================================================
   public:
// =======
//
// Type conversion operator
// -> number = int(mat) (assign the first matrix element)
// ======================================================
operator int ();                          // Overloaded Operator int
```

```
//
// Matrix index -> (i)
// ==================
int &operator () (int i);                       // Overloaded Operator ()
// Matrix index -> (i,j)
// ====================
int &operator () (int i, int j);                // Overloaded Operator ()
// Matrix index -> (i,j,k)
// ======================
int &operator () (int i, int j, int k);         // Overloaded Operator ()
// Matrix index -> (i,j,k,l)
// ========================
int &operator () (int i, int j, int k, int l);
//                                              // Overloaded Operator ()
//
// Matrix element
// -> mat.MATRIX_element(i,j,k,l) compare with mat(i,j,k,l)
// ===========================================================
int &element(int i=0, int j=0, int k=0, int l=0);
//
// Get the current matrix element and increment the element pointer
// -> mat++
// ===================================================================
// by explicitly calling
// -> mat.operator ++ (distance)
// the element pointer can be incremented by several elements,
// the element pointer is only changed by the constructor, destructor,
// and the operators '()', '--' !
// --------------------------------------------------------------------
int &operator ++ (int distance);                  // Overloaded Operator ++
//
// Get the current matrix element and decrement the element pointer
// -> mat--
// ====================================================================
// By explicitly calling
// -> mat.operator -- (distance)
// the element pointer can be decremented by several elements,
// the element pointer is only changed by the constructor, destructor,
// and the operators '()', '++' !
// --------------------------------------------------------------------
int &operator -- (int distance);                  // Overloaded Operator --
//
// Temporarily change matrix dimensions
// -> mat.dimension(i,j,k,l)
// ====================================
// the bundeling of indices (redimensioning) is actually needed
// for the matrix multiplication mat1*mat2,
// therefore, the redimensioning only applies
//                                   until the next matrix operation!
// afterwards, the original dimensions are restored !!!
// --------------------------------------------------------------------
intMATRIX &dimension(int ni=1, int nj=1, int nk=1, int nl=1);
//
// Determing matrix dimensions
```

```
// -> mat.determine_dimension(ni,nj,nk,nl)
// ========================================
// Result: dim=number of matrix elements (allocated memory)
   int dimension_ermitteln(int &ni, int &nj, int &nk, int &nl);
   int determine_dimension(int &ni, int &nj, int &nk, int &nl);
//
// Permanently change matrix dimensions
// -> mat.change_dimension(i,j,k,l)
// =================================
   intMATRIX &dimension_aendern(int ni=1, int nj=1, int nk=1,
                                                            int nl=1);
   intMATRIX &change_dimension (int ni=1, int nj=1, int nk=1,
                                                            int nl=1);
//
// Get number of rows/columns
// -> mat.rows_columns(ni,nj,nk,nl,dim)
// ==================================================================
// Result: number of rows/columns=sqrt(ni*nj*nk*nl)
//          for square matrices,
//          otherwise -1
   int zeilem_spalten(int &ni, int &nj, int &nk, int &nl);
   int rows_columns  (int &ni, int &nj, int &nk, int &nl);
//
// Extract a partial matrix
// -> ...=mat.partialmatrix_out(i, j, ni, nj, k, l, nk, nl)
// =========================================================
// partialmatrix_out retrieves consecutive matrix elements starting at
// index mat(i,j,k,l), that are stored in the submatrix(ni,nj,nk,nl)
// ------------------------------------------------------------------
   intMATRIX &teilmatrix_heraus(int i=0, int j=0, int ni=1, int nj=1,
                                int k=0, int l=0, int nk=1, int nl=1);
   intMATRIX &partialmatrix_out(int i=0, int j=0, int ni=1, int nj=1,
                                int k=0, int l=0, int nk=1, int nl=1);
//
// Insert partial matrix
// -> mat.partialmatrix_in(i, j, k, l)=...
// ========================================
// partialmatrix_in allows storing the partial matrix starting
//                                                  at index mat(i,j,k,l)
// -------------------------------------------------------------------
   intMATRIX &teilmatrix_hinein(int i=0, int j=0, int k=0, int l=0);
   intMATRIX &partialmatrix_in (int i=0, int j=0, int k=0, int l=0);
//
// Extract submatrix
// -> ...=mat.submatrix_out(i, ni, j, nj, k, nk, l, nl)
// ====================================================
// submatrix_out retrieves the submatrix starting
//                          at index mat(i,j,k,l) of size (ni,nj,nk,nl)
// ------------------------------------------------------------------
   intMATRIX &untermatrix_heraus(int i=0, int ni=1, int j=0, int nj=1,
                                 int k=0, int nk=1, int l=0, int nl=1)
                                 ;
   intMATRIX &submatrix_out      (int i=0, int ni=1, int j=0, int nj=1,
```

```
                                    int k=0, int nk=1, int l=0, int nl=1)
                                        ;
//
// Insert submatrix
// -> mat.submatrix_in(i, j, k, l)=...
// ====================================
// submatrix_in allows storing the submatrix starting
//                                                  at index mat(i,j,k,l)
// -------------------------------------------------------------------------
   intMATRIX &untermatrix_hinein(int i=0, int j=0, int k=0, int l=0);
   intMATRIX &submatrix_in       (int i=0, int j=0, int k=0, int l=0);
//
// Matrix output -> cout<<mat
// ==========================
friend ostream &operator << (ostream &s, intMATRIX &mat);
                                               // Overloaded Operator <<
//
// Matrix input -> cin>>mat
// ========================
friend istream &operator >> (istream &s, intMATRIX &mat);
                                               // Overloaded Operator >>
//
// Set all matrix elements -> mat =number
// ======================================
intMATRIX &operator  = (int number);          // Overloaded Operator =
// Set all matrix elements -> mat+=number
// ======================================
intMATRIX &operator += (int number);          // Overloaded Operator +=
// Set all matrix elements -> mat-=number
// ======================================
intMATRIX &operator -= (int number);          // Overloaded Operator -=
//
// Assign matrix -> mat1 =mat2 (copy matrix content only)
// ==========================
intMATRIX &operator  = (intMATRIX &mat2);   // Overloaded Operator  =
// Assign matrix -> mat1+=mat2 (copy matrix content only)
// ==========================
intMATRIX &operator += (intMATRIX &mat2);   // Overloaded Operator +=
// Assign matrix -> mat1-=mat2 (copy matrix content only)
// ==========================
intMATRIX &operator -= (intMATRIX &mat2);   // Overloaded Operator -=
//
// Positive sign -> +mat
// ====================
intMATRIX &operator + ();                      // Overloaded Operator +
//
// Addition with scalar -> mat+number
// ==================================
intMATRIX &operator + (int number);           // Overloaded Operator +
//
// Addition with scalar -> number+mat
// ==================================
friend intMATRIX &operator + (int number, intMATRIX &mat);
                                               // Overloaded Operator +
```

```
//
// Addition -> mat1+mat2
// =====================
intMATRIX &operator + (intMATRIX &mat2);     // Overloaded Operator +
//
// Negative sign -> -mat
// =====================
intMATRIX &operator - ();                    // Overloaded Operator -
//
// Subtraction with scalar -> mat-number
// =====================================
intMATRIX &operator - (int number);          // Overloaded Operator -
//
// Subtraction with scalar -> number-mat
// =====================================
friend intMATRIX &operator - (int number, intMATRIX &mat);
                                             // Overloaded Operator -
//
// Subtraction -> mat1-mat2
// ========================
intMATRIX &operator - (intMATRIX &mat2);     // Overloaded Operator -
//
// Multiplication with scalar -> mat*number
// ========================================
intMATRIX &operator * (int number);          // Overloaded Operator *
//
// Multiplication with scalar -> number*mat
// ========================================
friend intMATRIX &operator * (int number, intMATRIX &mat);
                                             // Overloaded Operator *
// Transpose matrix
// -> mat.transpose(index_0, index_1, index_2, index_3)
// ====================================================
// transpose stores all matrix elements around,
// index_0, index_1, index_2, index_3 indicate
//                                   which indices are swapped:
// e.g.,    mat.transpose(3,2,1,0)
//                        -> mat1(l3,k2,j1,i0) <= mat(i0,j1,k2,l3),
// default case: mat.transpose()
//                        -> mat1(k,l,i,j)     <= mat(i,j,k,l),
// 2D-matrices:  mat.transpose()
                          -> possible without input parameters
// -------------------------------------------------------------------
   intMATRIX &transponieren
       (int index_0=2, int index_1=3, int index_2=0, int index_3=1);
   intMATRIX &transpose
       (int index_0=2, int index_1=3, int index_2=0, int index_3=1);
};
// ===================================================================
```

3.4.1 Derived Classes e_MATRIX and delta_MATRIX

The class e_MATRIX [Meissner 2023] inherits all properties of the class MATRIX through the declaration 'class e_MATRIX : public MATRIX'.

The e_MATRIX class constructor can be used to create two- and three-dimensional e-matrices (2.2.4) and also ε-tensors (4.5.4), (4.5.5) with elements of type 'double'. Output is performed using the inherited operator '<<'.

Example:

```
e_MATRIX e1(2,2);                          // static matrix  e1

MATRIX& e2 = *new e_MATRIX(3,3,3);         // dynamic matrix e2
cout << e1 << e2;                          // print
delete& e2;                                // delete using destructor '~'
```

Declaration file (header file): e_Matrix.h

Listing 3.5: Derived Class e_MATRIX

```
// =============== e_MATRIX class ====================================
// Declaration file (header file)
class e_MATRIX : public MATRIX
// =============== Declaration of prototypes ======================
{
// e(i,j) or e(i,j,k) matrix
// Indexing of i, j, k starts at 0 !!!
// e-matrices are stored column-wise !!!
// -------------------------------------------------------------------
// Copyright: Univ.-Prof. Dr.-Ing. Udo F. Meissner, Erkrath/Darmstadt
// Advanced teaching and study program - Commercial use not permitted.
// The author assumes no liability for incompleteness and errors!
// Version:  2023
// Compiler: C++
// ===================================================================
//                      Public member functions
//                      =======================
   public:
// =======
//
// Constructor -> e_MATRIX e(ni,nj,factor)
// ========================================
// creates 2D e-matrices or epsilon tensors
// for epsilon tensors, factor = sqrt(g) or 1/sqrt(g) is required
// --------------------------------------------------------------
e_MATRIX(int ni,int nj,double factor = 1.0);
//
// Constructor -> e_MATRIX e(ni,nj,nk,factor)
// ===========================================
// creates 3D e-matrices or epsilon tensors
// for epsilon tensors, factor = sqrt(g) or 1/sqrt(g) is required
// --------------------------------------------------------------
```

```
e_MATRIX(int ni,int nj,int nk,double factor=1.0);
//
// Destructor -> delete &e
// ========================
~e_MATRIX();
};
// ==================================================================
```

The class delta_MATRIX [Meissner 2023] inherits all properties of the class MATRIX through the declaration 'class delta_MATRIX : public MATRIX'.

The delta_MATRIX class can be used to create two- to four-dimensional δ-matrices (2.2.2) with elements of type 'double' using the constructor. Output is performed using the inherited operator '<<'.

Example:

```
delta_MATRIX d1(2,2);                           // static matrix  d1

MATRIX& d2 = *new delta_MATRIX(3,3);            // dynamic matrix d2
cout << d2;                                     // print
delete& d2;                              // delete using destructor '~'
```

Declaration file (header file): delta_Matrix.h

Listing 3.6: Derived Class delta_MATRIX

```
// =============== delta_MATRIX class ===============================
// Declaration file (header file)
class delta_MATRIX : public MATRIX
// =============== Declaration of prototypes ========================
{
// delta(i,j) resp. delta(i,j,k,l) matrix
// Indexing of i, j, k starts at 0 !!!
// delta-matrices are stored column-wise !!!
// ------------------------------------------------------------------
// Copyright: Univ.-Prof. Dr.-Ing. Udo F. Meissner, Erkrath/Darmstadt
// Advanced teaching and study program - Commercial use not permitted.
// The author assumes no liability for incompleteness and errors!
// Version:  2023
// Compiler: C++
// ==================================================================
//                    Public member functions
//                    ========================
   public:
// =======
//
// Constructor -> delta_MATRIX delta(ni,nj, factor=1.0)
// ====================================================
// create delta-matrices of 2nd order
// 2D-matrix may be generated as diagonal matrix with factor
// ----------------------------------------------------------
delta_MATRIX(int ni,int nj, double factor=1.0);
//
```

```
// Constructor -> delta_MATRIX delta(ni,nj,nk,nl)
// ============================================
// create Kronecker symbol of 4th order
// 3D: e(i,j,s) * e(k,l,s)  for ni = nj = nk = nl = 3
// 2D: e(i,j)   * e(k,l)    for ni = nj = nk = nl = 2
// Grassmann identity: d(i,k ) * d(j,l) - d(i,l) * d(j,k)
// -------------------------------------------------------
delta_MATRIX(int ni,int nj,int nk,int nl);
//
// Destructor -> delete &delta
// ============================
~delta_MATRIX();
};
// ================================================================
```

3.4.2 Derived Classes VEKTOR and intVEKTOR

The class VEKTOR[1] [Meissner 2023] inherits all properties of the class MATRIX through the declaration 'class VEKTOR : public MATRIX'.

The two constructors of the VEKTOR class can be used to create vectors with elements of type 'double'. The vector elements may be preloaded using an array in notation of the programming language C. Input and output operations are performed using the inherited operators '>>' and '<<'.

Example:

```
double F8[8] = {1,2,3,4,5,6,7,8};  // preloaded C array F8
VEKTOR V1(F8, 8);                  // static vector V1, preloaded with F8

VEKTOR& V2 = *new VEKTOR(8);       // dynamic vector V2
        V2 = V1 * 2.0;             // assignment V2 := V1 * 2.0
cout << V1 << V2;                  // print
delete& V2;                        // delete using destructor '~'
```

Declaration file (header file): Vektor.h

Listing 3.7: Derived Class VEKTOR

```
// ================ VEKTOR class ======================================
// Declaration file (header file)
class VEKTOR : public MATRIX
// ================ Declaration of prototypes =======================
{
// double vector vek(i)
// Indexing of i starts at 0 !!!
// -----------------------------------------------------------------
// Copyright: Univ.-Prof. Dr.-Ing. Udo F. Meissner, Erkrath/Darmstadt
// Advanced teaching and study program - Commercial use not permitted.
// The author assumes no liability for incompleteness and errors!
// Version:  2023
// Compiler: C++
// =================================================================
//                     Public member functions
//                     ======================
   public:
// =======
//
// Constructor -> VEKTOR vek(ni)
// ===========================
// Create a vector with internal storage vek(1,ni)
// -----------------------------------------------
VEKTOR(int ni=1);
//
// Constructor -> VEKTOR vek(C_array,ni)
// =====================================
```

[1] engl. VECTOR

```
// Create a vector with internal storage vek(1,ni)
// C_array initialized using an initialization list
// -> C_array [] = { ... }
// ---------------------------------------------------
VEKTOR(double* C_array,int ni = 1);
//
// Destructor -> delete &vek
// ========================
~VEKTOR();
//
// Fill all vector elements -> vek =number
// ========================================
// provided again because the assignment operator =
//                                         cannot be inherited !!!
// -----------------------------------------------------------------
VEKTOR &operator  = (double number);        // Overloaded Operator =
//
// Assign matrix to vector -> vek =mat (copy matrix content only)
// ====================================
// provided again because the assignment operator =
//                                         cannot be inherited !!!
// -----------------------------------------------------------------
VEKTOR &operator  = (MATRIZEN &mat);        // Overloaded Operator =
//
// Scalar multiplication -> vek*number
// ====================================
MATRIZEN &operator * (double number);       // Overloaded Operator *
//
// Scalar product -> vek1*vek2 -> result = vek1(s) * vek2(s)
// ==========================================================
double operator * (VEKTOR &vek2);           // Overloaded Operator *
//
// Vector-matrix multiplication
// -> vek*mat -> c(k,l) = vek(s) * mat(k,s,l)
// ==========================================
MATRIZEN &operator * (MATRIZEN &mat);       // Overloaded Operator *
//
// Matrix-vector multiplication
// -> mat*vek -> c(i,j) = mat(i,s,j) * vek(s)
// ==========================================
// Using the inherited operator mat1 * mat2,
// because vek(s) is stored internally as vek(1,s)
// -----------------------------------------------------------------
//
// Search for a VALUE
// ==================
// Arguments:
// knonr: value to search for
// n    : length of the vector
//      : -1 automatically determine the length
// d    : step size for searching the vector
// Function result: relative starting index
//                  (0,1,..) when knonr is found
//                        -1 knonr was not found
```

```
    int wert_suchen (double knonr, int n=-1, int d=1);
    int search_value(double knonr, int n=-1, int d=1);
};
// ================================================================
```

The class intVEKTOR[1] [Meissner 2023] inherits all properties of the class intMATRIX through the declaration 'class intVEKTOR : public intMATRIX'.

The two constructors of the intVEKTOR class can be used to create vectors with elements of type 'int'. The vector elements may be preloaded using an array in notation of the programming language C. Input and output operations are performed using the inherited operators '>>' and '<<'.

Example:

```
int P4[4] = { 0, 1, 2, 3 };            // preloaded C array P4
intVEKTOR v1(P4, 4);            // static vector v1, preloaded with P4

intVEKTOR& v2 = *new intVEKTOR(10); // dynamic vector v2
           v2 = v1 * 4;             // assignment v2 := v1 * 4
cout << v1 << v2;                   // print
delete& v2;                         // delete using destructor '~'
```

Declaration file (header file): intVektor.h

Listing 3.8: Derived Class intVEKTOR

```
// =============== intVEKTOR class ===================================
// Declaration file (header file)
class intVEKTOR : public intMATRIX
// =============== Declaration of prototypes =========================
{
// int vector vek(i)
// Indexing of i starts at 0 !!!
// -------------------------------------------------------------------
// Copyright: Univ.-Prof. Dr.-Ing. Udo F. Meissner, Erkrath/Darmstadt
// Advanced teaching and study program - Commercial use not permitted.
// The author assumes no liability for incompleteness and errors!
// Version:  2023
// Compiler: C++
// ===================================================================
//                   Public member functions
//                   =======================
   public:
// =======
//
// Constructor -> intVEKTOR vek(ni)
// ================================
// Create a vector with internal storage vek(1,ni)
// ---------------------------------------------
intVEKTOR(int ni=1);
//
```

[1] engl. intVECTOR

```
// Constructor -> intVEKTOR vek(C_array,ni)
// ==========================================
// Create a vector with internal storage vek(1,ni)
// C_array initialized using an initialization list
// -> C_array [] = { ... }
// ------------------------------------------------
intVEKTOR(int* C_array, int ni = 1);
//
// Destructor -> delete &vek
// ========================
~intVEKTOR();
//
// Fill all vector elements -> vek =number
// ========================================
// provided again because the assignment operator =
//                                          cannot be inherited !!!
// ------------------------------------------------------------------
intVEKTOR &operator  = (int number);          // Overloaded Operator =
//
// Assign matrix to vector -> vek =mat (copy matrix content only)
// ===================================
// provided again because the assignment operator =
//                                          cannot be inherited !!!
// ------------------------------------------------------------------
intVEKTOR &operator  = (intMATRIX &mat);      // Overloaded Operator =
//
// Scalar multiplication -> vek*number
// ===================================
intMATRIX &operator * (int number);           // Overloaded Operator *
//
// Search for a VALUE
// =================
// Arguments:
// knonr: value to search for
// n    : length of the vector
//      : -1 automatically determine the length
// d    : step size for searching the vector
// Function result: relative starting index
//                  (0,1,..) when knonr is found
//                        -1 knonr was not found
   int wert_suchen (int knonr, int n=-1, int d=1);
   int search_value(int knonr, int n=-1, int d=1);
};
// ==================================================================
```

3.4.3 Derived Class qMATRIX for Matrix Division

With the matrix class qMATRIX, the solution of linear equation systems using the decomposition methods described in Section (2.3) and the determination of the inverse matrix (2.3.16) can be compactly summarized. For this purpose, it is necessary to capture the properties of the coefficient matrix within the constructor and store the staggered triangular matrices to process multiple right-hand sides as well as the generation of the inverse matrix. Moreover, the various methods of solving linear equation systems from Section (3.3.2) can be supplemented by declaring the class method 'Matrix Division' by its new memorable '/' operator:

$$\boxed{x_K = r_I / A_{IK} \quad \Leftrightarrow \quad \boldsymbol{x} = \boldsymbol{r} / \boldsymbol{A}} \tag{3.4.1}$$

resp. (no summation over I)

$$\boxed{Z_{KL} = \delta_{IL} / A_{IK} \quad \Leftrightarrow \quad \boldsymbol{Z} = \boldsymbol{I} / \boldsymbol{A} \;\; with \;\; \boldsymbol{Z} = \boldsymbol{A}^{-1}} \; . \tag{3.4.2}$$

Because matrix division is not defined in linear algebra, ***the overloaded operator '/' here can only be used procedurally*** for object-oriented programming purposes! It must also be emphasized that it should not be confused with the './' right division used in [MATLAB R2022a] for arrays.

The qMATRIX class [Meissner 2023] inherits all the properties of the base class MATRIX through the declaration 'class qMATRIX : public MATRIX'.

With the two constructors of the qMATRIX class, square coefficient matrices of type 'double' can be created. It is possible to predefine the matrix elements with numerical values using an array in the notation of the C programming language. Input and output are done through the inherited operators '>>' and '<<'.

Example:

```
double F9[9]                          // Pre-defined C array F9
                = {3.0,4.0,2.0, 1.0,3.0,2.0, -1.0,2.0,1.0};
qMatrix &A= *new qMatrix(F9,3,-1); // Create coefficient matrix A
delta_MATRIX I(3,3);                  // Create identity matrix I
MATRIX Z(3,3);                        // Create matrix Z

Z = I / A;                            // Calculate inverse matrix Z

I = A * Z.dimension(1,3,3);           // Test of A * Z
cout << A << K << I;                  // Print A, Z, and I
delete& A;                            // Delete A using destructor '~'
```

Declaration file (header file): qMatrix.h

Listing 3.9: Derived Class qMATRIX

```
// ======== qMATRIX class for square coefficient matrices =======
```

```
// Declaration file (header file)
class qMATRIX : public MATRIX
{
// ====== Declaration of Prototypes ==============================
// Coefficient matrix MAT(mk,mk) is stored column-wise!
// Indexing for mk starts from 0 !!!
// cf. U. Breymann: C++. Carl Hanser Verlag (1993), p.186 ff
// ---------------------------------------------------------------
// Copyright: Univ.-Prof. Dr.-Ing. Udo F. Meissner, Erkrath/Darmstadt
// Advanced teaching and study program - Commercial use not permitted.
// The author assumes no liability for incompleteness and errors!
// Version:  2023
// Compiler: C++
// ===============================================================
// Public Member Functions
// ======================
   public:
// =======
// Constructor -> qMATRIX MAT(mk,symdef)
// ==================================
// mk, mk: Dimensions of rows / columns of the 2D-matrix
// symdef =  1   symmetric positive-definite coefficient matrix
//        =  0   symmetric                   coefficient matrix
//        = -1 unsymmetric                   coefficient matrix
// ---------------------------------------------------------------
qMATRIX(int mk, int symdef = -1);
//
// Constructor -> qMATRIX MAT(C_array,mk,symdef)
// ==========================================
//C_array pre-initialized using an initialization list
// -> C_array [] = { ... }
// mk: Dimensions of rows / columns of the 2D-matrix
// symdef =  1   symmetric positive-definite coefficient matrix
//        =  0   symmetric                   coefficient matrix
//        = -1 unsymmetric                   coefficient matrix
// ---------------------------------------------------------------
qMATRIX(double* C_array = NULL, int mk = 1, int symdef = -1);
//
// Destructor -> delete &mat
// =======================
~qMATRIX();
//
//Public Member Functions (with temporary matrices)
//==============================================
   public:
// =======
//
// Assign matrix -> mat1 = mat2 (only copy matrix content)
// ====================================================
// Defined again here because the assignment operator =
// cannot be inherited !!!
// ---------------------------------------------------------------
qMATRIX& operator  = (MATRIX& mat2);  // Overloaded operator =
//
```

```
// 'Matrix Division' matx = matr / MATk
// -> matx(s,n) = matr(i,n) / MATk(i,s)
// ================================================================
// Solve equation system -> MATk(i,s) * matx(s,n) = matr(i,n)
// The staggered form of MATk is stored internally
// to process additional right-hand sides matr
// ----------------------------------------------------------------
MATRIZEN &operator / (qMATRIX &MATk); // Overloaded operator /
//
// Output calculated determinant
// -> symdef = MATk.get_determinant (det)
// ======================================
// symdef = 1 / 0 / -1 for
// Cholesky / Gauss (symmetric) / Gauss (unsymmetric)
// ----------------------------------------------------------------
   int& gib_determinante(double& det);
   int& get_determinant (double& det);
};
// ================================================================
```

References

Breymann 1993. Breymann, U.: C++ - Eine Einführung. 1. Aufl., Carl Hanser, München (1993)

Breymann 1997. Breymann, U.: C++ - Eine Einführung. 4. Aufl., Carl Hanser, München (1997)

Coad/Yourdon 1990. Coad, P.; Yourdon, E.: Object-Oriented Analysis. Prentice Hall, Englewood Cliffs (1990)

Hartmann (Hrsg.) 2000. Hartmann, D.: Objektorientierte Modellierung in Planung und Konstruktion - DFG-Bericht, Wiley-VCH, Weinheim (2000)

MATLAB R2022a. MathWorks: Produktfamilie (2022)

Meissner 2023. Meißner, U. F.: Object-oriented Tensor and Matrix Classes (2023). `https://rebrand.ly/Matrix_Classes`

Rumbaugh/Blaha/Premerlani/Eddy/Lorensen 1991. Rumbaugh, J.; Blaha, M.; Premerlani, W.; Eddy, F.; Lorensen, W.: Object-Oriented Modeling and Design. Prentice Hall, Englewood Cliffs (1991)

Chapter 4
Fundamentals of Tensor Calculus

Abstract

Tensor calculus, with its consistent formulation of invariants and transformations, has acquired a high significance in engineering over the years. Thus, early works on the mechanics of load-bearing structures, such as [Green/Zerna 1954], which renewed the theory's formulation, and mathematical compendia on tensor calculus in index notation, such as [Duschek/Hochrainer 1968], can serve as a basis at this point. In this chapter, we recapitulate the most important basics of tensor algebra in this context, especially to make the approach of object-oriented matrix calculus comprehensible and to provide the ability to apply the presented methodology consistently. This is because the index notation allows the arithmetic operations of tensor calculus to be seamlessly transformed into the practical numeric aspects of programming tools using the new object-oriented classes and methods of matrix calculus. This capability is demonstrated with the tensor class BASIS and its application in the numeric function METRIC.

The use of tensor calculus [Klingbeil 1966] has significantly contributed in the past to, on the one hand, establishing a consistent foundation for describing mechanical systems [Green/Zerna 1954], and on the other hand, carrying out unambiguous theorems for determining invariants and performing transformations between different reference systems [Duschek/Hochrainer 1968]. For a long time, the symbolic tensor notation was preferred, akin to traditional matrix calculus, with which the theory could be compactly formulated and written down. The practical and useful aspects of numerical software methods were hardly considered. It is only through the consistent formulation of all tensor quantities and relationships in index notation that tensor calculus - as a counterpart to matrix calculus - can be directly translated into numerical algorithms that may be intuitively implemented into corresponding structures of object-oriented software methods.

The following considerations aim to provide a consistent contribution to transform tensor quantities into corresponding matrix schemes and then translate them into numerical algorithms. In Chapter 2, the methods of matrix calculus were already presented, and their implementation in object-oriented software tools was explained in Chapter 3. Now in the following, the known methods of tensor algebra will be prepared accordingly and made accessible for numerical processing.

U. F. Meissner, *Tensor Calculus with Object-Oriented Matrices for Numerical Methods in Mechanics and Engineering*, https://doi.org/10.1007/978-3-031-59302-4_4

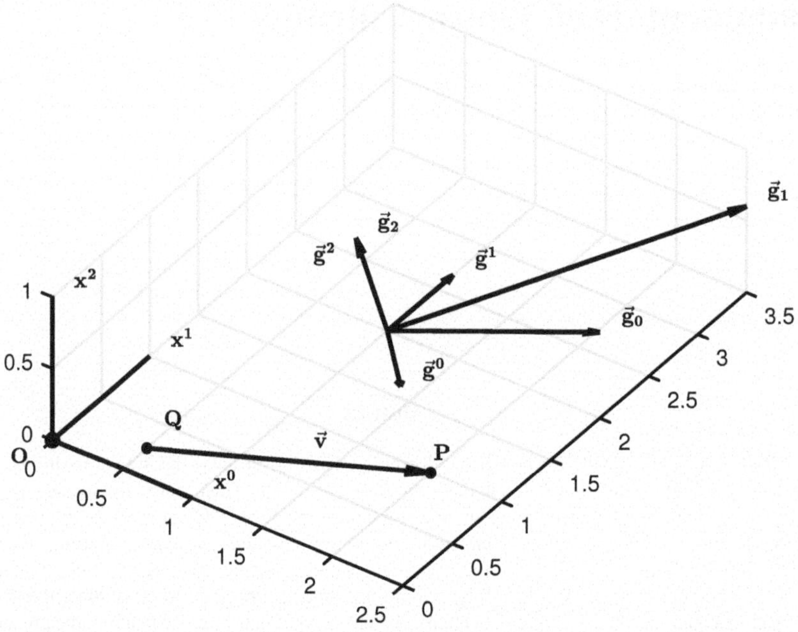

Fig. 4.1: Cartesian coordinates x^i, covariant and contravariant basis vectors $\mathbf{g}_\alpha$ and $\mathbf{g}^\alpha$

4.1 Tensor Algebra in Cartesian Reference Systems

In a ***Cartesian coordinate system***, one considers a system of three scaled lines in Euclidean space $\mathbb{R}^3$, all originating from the same point, with their normalized axes forming right angles to each other (see Fig. 4.1).

Each point $\boldsymbol{P}$ is uniquely associated with a coordinate triple $x^i{}_P = (x^0\, x^1\, x^2)_P$ by which a position vector can be constructed from the coordinate origin $\boldsymbol{O}$ to point $\boldsymbol{P}$. With the unit vectors

$$\overline{\boldsymbol{P_0O}} := \mathbf{e}_0 := (1\,0\,0)_{P_0} - (0\,0\,0)_O = e^i{}_0$$

$$\overline{\boldsymbol{P_1O}} := \mathbf{e}_1 := (0\,1\,0)_{P_1} - (0\,0\,0)_O = e^i{}_1$$

$$\overline{\boldsymbol{P_2O}} := \mathbf{e}_2 := (0\,0\,1)_{P_2} - (0\,0\,0)_O = e^i{}_2$$

the three ***basis vectors*** $\mathbf{e}_p$ resp. $\mathbf{e}^p$

$$\boxed{\mathbf{e}_p = \mathbf{e}^p := \begin{pmatrix} 1\,0\,0 \\ 0\,1\,0 \\ 0\,0\,1 \end{pmatrix} \quad \Leftrightarrow \quad e^i{}_p = e_i{}^p = \delta_{ip}} \quad \textit{for } i, p \in \mathbb{N}\,[0, \leq 2] \tag{4.1.1}$$

are formed at the coordinate origin, serving the purpose of measuring the point space as units.

With the introduction of the scalar product, the ***actual Euclidean vector space*** $\mathbb{V}^3$ is then spanned within the point space

$$\boxed{e^i{}_p * e^i{}_q = \delta_{pq} \quad \Leftrightarrow \quad \mathbf{e}_p * \mathbf{e}_q = \delta_{pq}}, \tag{4.1.2}$$

with its basis vectors orthogonal to each other as the δ-matrix displays.

Between two points $\boldsymbol{Q}$ and $\boldsymbol{P}$ in $\mathbb{V}^3$, we can introduce exactly one connecting vector $\bar{v}$ (see Fig. 4.1) that maps point $\boldsymbol{Q}$ to point $\boldsymbol{P}$:

$$\boldsymbol{P} := \boldsymbol{Q} + \bar{v}\,.$$

As shown above, the components of vector $\bar{v}$ then result from the coordinate differences Δx^p between $\boldsymbol{P}$ *and* $\boldsymbol{Q}$

$$v^p = (\Delta x^p)_{PQ} = (x^p)_P - (x^p)_Q$$

and their orientation refers to the basis vectors (4.1.1)

$$\overline{\boldsymbol{PQ}} := \mathbf{v} = v^0 * \mathbf{e}_0 + v^1 * \mathbf{e}_1 + v^2 * \mathbf{e}_2 = v^p * \mathbf{e}_p\,.$$

A ***vector*** $\boldsymbol{v}$ is always a directed line segment, characterized by the ***basis*** $\mathbf{e}_p$ and its ***components*** v^p (cf. Fig. 4.2)

$$\boxed{v^i = v^p * e^i{}_p \quad \Leftrightarrow \quad \mathbf{v} = v^p * \mathbf{e}_p}\,. \tag{4.1.3}$$

It is important to note that the vector is no longer tied to the coordinates of points $\boldsymbol{P}$ and $\boldsymbol{Q}$, but is solely determined by the coordinate difference $(\Delta x^p)_{PQ}$. Therefore, it is called ***transient*** in its direction, and its properties are independent of these locations.

When taking the scalar product of two vectors $\mathbf{v}$ *and* $\mathbf{u}$, with respect to (4.1.3), the result is

$$v^p * e^i{}_p * e^i{}_q * u^q = v^p * \delta_{pq} * u^q = v^p * u_p\,.$$

For $\mathbf{v} = \mathbf{u}$, this yields the ***invariant of vector* v**

$$\boxed{I_0 = v^p * v_p \quad \Leftrightarrow \quad I_0 = \mathbf{v} * \mathbf{v} = |\mathbf{v}|^2} \tag{4.1.4}$$

with the vector length $|\mathbf{v}|$. As to notice, this invariant is independent of the reference system of the basis vectors.

For $\mathbf{v} \neq \mathbf{u}$, we get the familiar relationship for the ***angle* $\boldsymbol{\Phi}$ *between two vectors***

$$\boxed{|v^p| * |u_p| * cos(\Phi) = v^p * u_p \quad \Leftrightarrow \quad |\mathbf{v}| * |\mathbf{u}| * cos(\Phi) = \mathbf{v} * \mathbf{u}} \,. \tag{4.1.5}$$

Thus, it is possible to measure lengths and angles in $\mathbb{V}^3$ and perform a mathematically precise survey of the visual space, which is one of the fundamental tasks of geodesy.

With the scalar product (4.1.5), the vector components v^i represent the projection of $\mathbf{v}$ onto the respective basis vectors $\mathbf{e}_p$

$$\boxed{|v^i| * cos(\Phi^p) = v^i * e_i{}^p \quad \Leftrightarrow \quad |\mathbf{v}| * cos(\Phi^p) = \mathbf{v} * \mathbf{e}_p} \,. \tag{4.1.6}$$

4.2 Tensor Algebra in Skewed Reference Systems

The reference system introduced in the previous section, consisting of orthonormalized basis vectors, will now be generalized by spanning a new vector space in $\mathbb{R}^3$ with arbitrarily long basis vectors, which no longer need to be orthogonal to each other. This ultimately allows to examine multi-level tensors and to perform transformations between different reference systems.

First, using (4.1.3), a new system of three ***covariant basis vectors*** $g^i{}_\alpha$ resp. $\mathbf{g}_\alpha$ (cf. Fig. 4.2) is defined, which should be linearly independent of each other for arbitrary $\lambda_\alpha \neq 0$ with $\lambda_\alpha * \mathbf{g}_\alpha \neq \mathbf{0}$:

$$g^i{}_0 = g^p{}_0 * e^i{}_p \quad bzw. \quad \mathbf{g}_0 = g^p{}_0 * \mathbf{e}_p$$

$$g^i{}_1 = g^p{}_1 * e^i{}_p \quad bzw. \quad \mathbf{g}_1 = g^p{}_1 * \mathbf{e}_p$$

$$g^i{}_2 = g^p{}_2 * e^i{}_p \quad bzw. \quad \mathbf{g}_2 = g^p{}_2 * \mathbf{e}_p$$

$$\boxed{g^i{}_\alpha = g^p{}_\alpha * e^i{}_p \quad \Leftrightarrow \quad \mathbf{g}_\alpha = g^p{}_\alpha * \mathbf{e}_p} \quad for\ \alpha \in \mathbb{N}\ [0, \leq 2]\ . \tag{4.2.1}$$

With α, the concrete dimensionality of the space $\mathbb{V}^{\alpha+1}$, as one-, two- or three-dimensional, embedded in $\mathbb{R}^3$ can initially remain open.

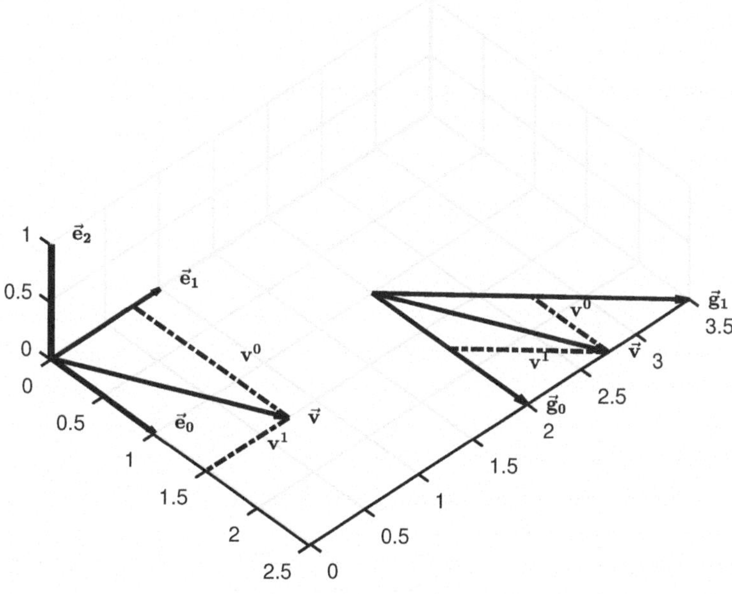

Fig. 4.2: Covariant $\mathbf{g}_\alpha$ basis vectors with contravariant v^α vector components

Convention for Tensor Notation

To distinguish between indices referring to the ***Cartesian basis vectors*** $\mathbf{e}_i$ or coordinates x^i, ***Latin indices*** (such as i, k, p, q) will always be used in the following.

Indices that refer to the ***skewed basis vectors*** $\mathbf{g}_\alpha$ or associated coordinates Θ^α will be named using ***Greek letters*** (such as α, β, γ, δ).

When it comes to tensors - in addition to the notation for matrices - a distinction is made between
- ***covariant tensors*** with lower indices (such as g_α, $T_{\alpha\beta}$),
- ***contravariant tensors*** with upper indices (such as g^α, $T^{\alpha\beta}$),
- ***mixed-variant*** resp. ***mixed tensors*** with both upper and lower indices (such as $T^\alpha{}_\beta$ or $T_\alpha{}^\beta$).

It is important to note that the quantities $T_{\alpha\beta} \neq T^{\alpha\beta} \neq T^\alpha{}_\beta$ are distinct!

However, when it comes to tensor or vector components represented in the Cartesian system, the distinction between covariant and contravariant does not matter, because the basis vectors $\mathbf{e}_p = \mathbf{e}^p$, as defined in Section (4.1.1), are identical.

Using (4.2.1), we can determine the metric coefficients of the symmetric ***covariant metric***

$$\boxed{g_{\alpha\beta} = g^i{}_\alpha * g^i{}_\beta \quad \Leftrightarrow \quad \mathbf{g} = \mathbf{g}_\alpha * \mathbf{g}_\beta}, \tag{4.2.2}$$

whose diagonal elements g_{00}, g_{11}, g_{22}, according to (4.1.4), represent the length squares $|\mathbf{g}_\alpha|^2$ of the basis vectors, and whose off-diagonal elements $g_{\alpha\beta}$, according to (4.1.5), represent the angles $|\mathbf{g}_\alpha| * |\mathbf{g}_\beta| * \cos(\phi_{\alpha\beta})$ between the basis vectors.

The ***determinant of the covariant metric***

$$\boxed{g = det(g_{\alpha\beta}) > 0} \tag{4.2.3}$$

is positive due to the linear independence of the basis vectors $\mathbf{g}_\alpha$, and it is calculated, as shown in Section (2.2.9):

$$g = det(g^i{}_\alpha) * det(g^i{}_\beta) = [e_{ikl} * g^i{}_0 * g^k{}_1 * g^l{}_2]^2$$

- in the one-dimensional case $\mathbb{R}^1$ according to (4.1.4) ***the length*** $\sqrt{g} = |\mathbf{g}_0|$ of the basis vector,
- in the two-dimensional case $\mathbb{R}^2$ according to (2.2.10) or with the cross product (2.2.7) ***the area*** $\sqrt{g} = |\mathbf{g}_0 \times \mathbf{g}_1|$ spanned by the two basis vectors,
- in the three-dimensional case $\mathbb{R}^3$ according to (2.2.11) or with the scalar triple product (2.2.8)

the volume $\sqrt{g} = (\mathbf{g}_0 \times \mathbf{g}_1) * \mathbf{g}_2$ spanned by the three basis vectors, as displayed in Fig. 4.3.

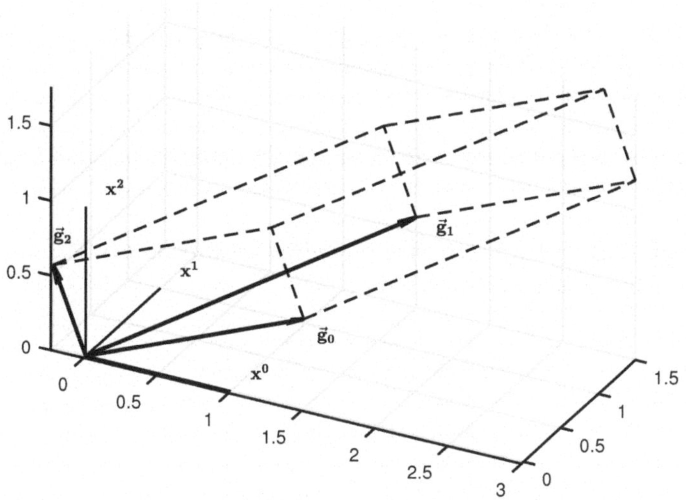

Fig. 4.3: Volume form $\sqrt{g}$ of covariant basis vectors $\mathbf{g}_\alpha$

The ***contravariant metric*** is determined as the inverse of the covariant metric

$$\boxed{g^{\alpha\beta} * g_{\beta\gamma} = \delta^\alpha{}_\gamma \quad \Leftrightarrow \quad \mathbf{g}^{-1} * \mathbf{g} = \boldsymbol{I}}, \tag{4.2.4}$$

whose diagonal elements g^{00}, g^{11}, g^{22} now represent the length squares $|\mathbf{g}^\alpha|^2$ of the contravariant basis vectors, and whose off-diagonal elements $g^{\alpha\beta}$ represent the angles $|\mathbf{g}^\alpha| * |\mathbf{g}^\beta| * \cos(\phi^{\alpha\,\beta})$ between the contravariant basis vectors.

Its determinant is the reciprocal value

$$\boxed{det(g^{\alpha\beta}) = 1/g = 1/det(g_{\alpha\beta}) > 0} \tag{4.2.5}$$

of the covariant metric.

The associated ***contravariant basis vectors*** (cf. Fig. 4.4) are obtained from the covariant basis vectors using the contravariant metric

$$\boxed{g_i{}^\alpha = g^{\alpha\beta} * g^i{}_\beta \quad \Leftrightarrow \quad \mathbf{g}^\alpha = g^{\alpha\beta} * \mathbf{g}_\beta}\,. \tag{4.2.6}$$

Because of (4.2.2), it holds

$$g_i{}^{\alpha} * g^i{}_{\gamma} = g^{\alpha\beta} * g^i{}_{\beta} * g^i{}_{\gamma} = g^{\alpha\beta} * g_{\beta\gamma} = \delta^{\alpha}{}_{\gamma} ,$$

demonstrating that the ***mixed-variant δ-metric***

$$\boxed{g^{\alpha}{}_{\gamma} = g_i{}^{\alpha} * g^i{}_{\gamma} = \delta^{\alpha}{}_{\gamma} \quad \Leftrightarrow \quad \mathbf{g}^{\alpha} * \mathbf{g}_{\gamma} = \boldsymbol{I}} \tag{4.2.7}$$

corresponds to the identity matrix.

Contravariant and covariant basis vectors are therefore orthogonal to each other (cf. Fig. 4.1).

Therefore, this relationship ***transforms the mixed identity matrix*** $\boldsymbol{\delta^i{}_p} = e^i{}_p$ of the orthogonal reference system according to (4.1.1)

$$g_k{}^{\alpha} * \delta^k{}_i * g^i{}_{\gamma} = g_i{}^{\alpha} * g^i{}_{\gamma} = \delta^{\alpha}{}_{\gamma}$$

and vice versa (4.2.8)

$$\delta^k{}_i = g^k{}_{\alpha} * \delta^{\alpha}{}_{\gamma} * g_i{}^{\gamma} = g^k{}_{\gamma} * g_i{}^{\gamma}$$

into the mixed-variant δ-metric of the skewed basis system.

4.3 Transformations and Invariants of First-Order Tensors

In Section 4.1, the concept of a vector was introduced in Cartesian coordinates using (4.1.3), and in Section 4.2, it was extended to skew basis vectors. The following will explore the rules for transforming its skewed vector components into orthogonal base systems and vice versa.

For this purpose, following (4.1.3) the ***definition of a vector*** is provided

$$\boxed{v^p * \mathbf{e}_p = \mathbf{v} = v^\alpha * \mathbf{g}_\alpha}\,, \tag{4.3.1}$$

and with (4.2.1), the covariant basis is substituted in this equation:

$$v^p * \mathbf{e}_p = \mathbf{v} = v^\alpha * g^p{}_\alpha * \mathbf{e}_p \,.$$

The result is the desired transformation

$$v^i = v^\alpha * g^i{}_\alpha \,.$$

By multiplying with the contravariant basis $g_i{}^\beta$ and using (4.2.7) this result

$$g_i{}^\beta * v^i = g_i{}^\beta * g^i{}_\alpha * v^\alpha = \delta^\beta{}_\alpha * v^\alpha$$

can be reversed

$$g_i{}^\beta * v^i = v^\beta \,.$$

Raising and lowering indices yields the corresponding results for covariant vector components.

In summary, the ***transformation equations*** for ***covariant and contravariant vector components*** (cf. Fig. 4.2 and Fig. 4.4) are:

$$\boxed{\begin{array}{ll} v^\alpha = g_i{}^\alpha * v^i & v_\alpha = g^i{}_\alpha * v_i \\[2ex] v^i = g^i{}_\alpha * v^\alpha & v_i = g_i{}^\alpha * v_\alpha \end{array}}\,. \tag{4.3.2}$$

Definition and Transformation of First-Order Tensors

These transformation equations serve together also as the definition for tensors; because all vector quantities that transform according to these rules are ***first-order tensors***.

Note the strict opposite orientation of indices. Especially, the summation indices must be oppositely arranged; otherwise, severe errors in formulation may arise.

Exceptions to this can generally only occur with Latin indices, which are related to the Cartesian basis. In this particular case, there exists no strict distinction between covariant and contravariant basis vectors, because the covariant unit metric $\mathbf{e}_i = \delta_{ik} * \mathbf{e}^k$ or the contravariant unit metric $\mathbf{e}^i = \delta^{ik} * \mathbf{e}_k$ are used.

With the metrics (4.2.2) and (4.2.4), covariant and contravariant tensor components can also be converted into each other. For example, using the first transformation equation and replacing v^i by v_i

$$v^\alpha = g_i{}^\alpha * v^i = g_i{}^\alpha * g_i{}^\beta * v_\beta \,,$$

and using the contravariant metric, the desired result

$$v^\alpha = g^{\alpha\beta} * v_\beta$$

is obtained.

The further ***transformations between covariant and contravariant tensor components*** follow accordingly

$$\boxed{\begin{array}{ll} v^\alpha = g^{\alpha\beta} * v_\beta & v_\alpha = g_{\alpha\beta} * v^\beta \\ v^\alpha = \delta^\alpha{}_\beta * v^\beta & v_\alpha = \delta_\alpha{}^\beta * v_\beta \end{array}}\,, \tag{4.3.3}$$

where by use of the mixed-variant δ-metric, only the indices are exchanged.

Through the transformation Eq's. (4.3.2), now the characteristic generation of tensor invariants can be established. For this purpose, one multiplies

$$v^i * v_i = v^\alpha * g^i{}_\alpha * g_i{}^\beta * v_\beta = v^\alpha * \delta_\alpha{}^\beta * v_\beta \,,$$

and obtains, corresponding to (4.1.4), the ***invariant of a first-order tensor***

$$\boxed{\mathrm{I}_0 = |\mathbf{v}|^2 = v^\alpha * v_\alpha = v_\beta * v^\beta} \tag{4.3.4}$$

as a scalar quantity I_0, being independent of a reference system.

Invariants and Definition of First-Order Tensors

For a first-order tensor, an invariant can be formed by pairwise contraction of its covariant and contravariant indices, which scalar value is independent of the reference system of basis vectors.

Since this characteristic property is directly derived from the transformation rules (4.3.2), it can also be used conversely as the definition of a ***tensor of the first order***, from which the rules of the transformations follow directly.

Note here the strict opposite up-and-down position of indices.

It holds that products like $v^\alpha * v^\alpha \neq v_\beta * v_\beta$ do not result in invariants!

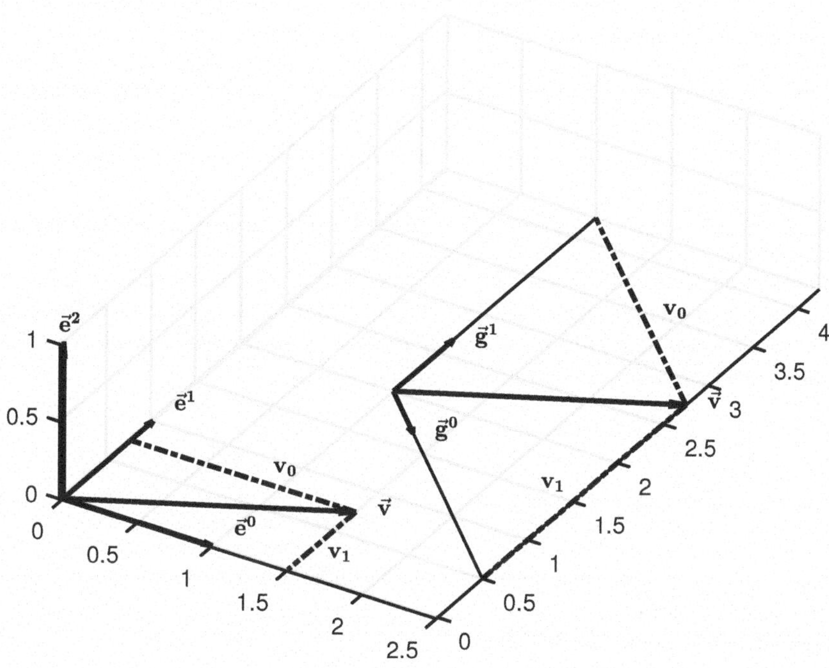

Fig. 4.4: Contravariant $\mathbf{g}^\alpha$ basis vectors with covariant v_α vector components

4.4 Tensors of Higher Order

Tensors of higher order represent an extension of the vector concept introduced in (4.3.1).

The vector space $\mathbb{R}^3$ with $\lambda_\alpha * \mathbf{g}_\alpha \neq 0$ linearly independent basis vectors is extended to dimension $\mathbb{R}^{3\times n}$ with n covariant and contravariant vectors. A tensor of nth order $\mathbf{T}^{(\mathbf{n})}$ has n covariant resp. contravariant associated components that form the projection onto this tensor basis

$$\boxed{T^{pq\,..}{}_{rs\,..} * \mathbf{e}_p * \mathbf{e}_q * .. \mathbf{e}^r * \mathbf{e}^s .. = \mathbf{T}^{(\mathbf{n})} = \mathbf{T}^{\alpha\beta\,..}{}_{\lambda\mu\,..} * \mathbf{g}_\alpha * \mathbf{g}_\beta * .. \mathbf{g}^\lambda * \mathbf{g}^\mu ..} \quad . \tag{4.4.1}$$

For the metric space, the scalar product holds, and as before, it is a characteristic property that the tensor satisfies the transformation rules for each index when changing reference systems, as given by (4.3.2) and (4.3.3).

Thus, any tensor of nth order can be projected onto a tensor of $(n-1)$th order through a linear mapping using a 1st-order tensor:

$$\begin{array}{lllll}
T^p * T_p & = T & = T^\lambda * T_\lambda & \Leftrightarrow\ T^{(1)} * T = T^{(0)} & \text{tensor 0th order (scalar)} \\
T^p{}_r * T^r & = T^p & = T^p{}_\lambda * T^\lambda & \Leftrightarrow\ \mathbf{T}^{(2)} * \mathbf{T} = \mathbf{T}^{(1)} & \text{tensor 1st order (vector)} \\
T^p{}_r{}^q * T^r & = T^{pq} & = T^p{}_\lambda{}^q * T^\lambda & \Leftrightarrow\ \mathbf{T}^{(3)} * \mathbf{T} = \mathbf{T}^{(2)} & \text{tensor 2nd order} \\
\ldots & & & & \ldots \\
 & & & \Leftrightarrow\ \mathbf{T}^{(\mathbf{n})} * \mathbf{T} = \mathbf{T}^{(\mathbf{n-1})} & \text{tensor (n-1)th order}
\end{array}$$

As the example shows, the mappings are independent of the chosen basis reference system due to the transformation properties. Furthermore, from the concatenation of the mappings, it can also be seen that a ***tensor of higher order transforms like a product of 1st-order tensors***, with each index being transformed individually

$$\boxed{\begin{array}{l}
T^{pq\,\ldots}{}_{rs\,\ldots} = T^{\alpha\beta\,\ldots}{}_{\lambda\mu\,\ldots} * g^p{}_\alpha * g^q{}_\beta * \ldots g_r{}^\lambda * g_s{}^\mu \ldots \\
T^{\alpha\beta\,\ldots}{}_{\lambda\mu\,\ldots} = T^{pq\,\ldots}{}_{rs\,\ldots} * g_p{}^\alpha * g_q{}^\beta * \ldots g^r{}_\lambda * g^s{}_\mu \ldots
\end{array}} \quad . \tag{4.4.2}$$

As in Section 4.3, the invariants for tensors of higher order can now be determined by contraction of indices.
Using (4.4.2), the ***2nd-order tensor***

$$T^p{}_q = T^\alpha{}_\beta * g^p{}_\alpha * g_q{}^\beta$$

can first be considered as an example. Using $g^p{}_\alpha * g_p{}^\beta = \delta_\alpha{}^\beta$, and by contraction of indices p and q,
the ***1st invariant*** of this tensor

$$I_{11} = T^p{}_p = T^\alpha{}_\alpha \quad \Leftrightarrow \quad I_{11} = \mathrm{tr}(\mathbf{T})$$

is obtained immediately.

Next, $T^p{}_q$ is multiplied with

$$T^r{}_s = T^\lambda{}_\mu * g^r{}_\lambda * g_s{}^\mu$$

and the result is

$$T^p{}_q * T^r{}_s = T^\alpha{}_\beta * g^p{}_\alpha * g_q{}^\beta * g^r{}_\lambda * g_s{}^\mu * T^\lambda{}_\mu .$$

By contraction of the covariant and contravariant indices p and q with s and r

$$T^p{}_q * T^q{}_p = T^\alpha{}_\beta * g^p{}_\alpha * g^q{}_\lambda * g_q{}^\beta * g_p{}^\mu * T^\lambda{}_\mu = T^\alpha{}_\beta * \delta_\alpha{}^\mu * \delta_\lambda{}^\beta * T^\lambda{}_\mu ,$$

the ***2nd invariant***

$$I_{12} = T^p{}_q * T^q{}_p = T^\mu{}_\beta * T^\beta{}_\mu \quad \Leftrightarrow \quad I_{12} = \mathrm{tr}(\mathbf{T} * \mathbf{T})$$

is found.

Switching the indices again yields the ***3rd invariant***

$$I_{13} = T_q{}^p * T^q{}_p = T_\beta{}^\alpha * T^\beta{}_\alpha \quad \Leftrightarrow \quad I_{13} = \mathrm{tr}(\mathbf{T}^T * \mathbf{T}) .$$

Another invariant can be obtained by further multiplication of

$$T^p{}_q * T^q{}_s = T^\alpha{}_\beta * g^p{}_\alpha * g_q{}^\beta * g^q{}_\lambda * g_s{}^\mu * T^\lambda{}_\mu = T^\alpha{}_\beta * g^p{}_\alpha * g_s{}^\mu * T^\beta{}_\mu$$

with

$$T^s{}_p = T^\gamma{}_\delta * g^s{}_\gamma * g_p{}^\delta ,$$

leading with

$$\begin{aligned} T^p{}_q * T^q{}_s * T^s{}_p &= T^\alpha{}_\beta * g^p{}_\alpha * g_p{}^\delta * T^\gamma{}_\delta * g_s{}^\mu * g^s{}_\gamma * T^\beta{}_\mu \\ &= T^\alpha{}_\beta * \delta^\mu{}_\gamma * T^\beta{}_\mu * \delta_\alpha{}^\delta * T^\gamma{}_\delta \end{aligned}$$

to the ***4th invariant***

$$I_{14} = T^p{}_q * T^q{}_s * T^s{}_p = T^\alpha{}_\beta * T^\beta{}_\gamma * T^\gamma{}_\alpha \quad \Leftrightarrow \quad I_{14} = tr(\boldsymbol{T} * \boldsymbol{T} * \boldsymbol{T}) .$$

By swapping the indices, we obtain the ***5th invariant***

$$I_{15} = T^p{}_q * T^q{}_s * T_p{}^s = T^\alpha{}_\beta * T^\beta{}_\gamma * T_\alpha{}^\gamma \quad \Leftrightarrow \quad I_{15} = tr(\boldsymbol{T} * \boldsymbol{T} * \boldsymbol{T}^T) .$$

In this compilation, the five ***invariants of a 2nd-order tensor*** are represented as follows:

$$
\begin{array}{ll}
I_{11} = T^p{}_p = T^\alpha{}_\alpha & \Leftrightarrow \; I_{11} = tr(\boldsymbol{T}) \\
I_{12} = T^p{}_q * T^q{}_p = T^\alpha{}_\beta * T^\beta{}_\alpha & \Leftrightarrow \; I_{12} = tr(\boldsymbol{T} * \boldsymbol{T}) \\
I_{13} = T_q{}^p * T^q{}_p = T_\beta{}^\alpha * T^\beta{}_\alpha & \Leftrightarrow \; I_{13} = tr(\boldsymbol{T}^T * \boldsymbol{T}) \\
I_{14} = T^p{}_q * T^q{}_s * T^s{}_p = T^\alpha{}_\beta * T^\beta{}_\gamma * T^\gamma{}_\alpha & \Leftrightarrow \; I_{14} = tr(\boldsymbol{T} * \boldsymbol{T} * \boldsymbol{T}) \\
I_{15} = T^p{}_q * T^q{}_s * T_p{}^s = T^\alpha{}_\beta * T^\beta{}_\gamma * T_\alpha{}^\gamma & \Leftrightarrow \; I_{15} = tr(\boldsymbol{T} * \boldsymbol{T} * \boldsymbol{T}^T) \\
 & \\
\textit{In the case of tensor symmetry} & \\
T^p{}_q = T_q{}^p \;\; \textit{resp.} \;\; T^\alpha{}_\beta = T_\beta{}^\alpha & \Leftrightarrow \; \boldsymbol{T} = \boldsymbol{T}^T \\
\textit{are the invariants} & \\
I_{12} = I_{13} \textit{ and } I_{14} = I_{15} \textit{ identical.} & \\
 & \\
\textit{If the tensor with} & \\
T^p{}_q = -T_q{}^p \;\; \textit{resp.} \;\; T^\alpha{}_\beta = -T_\beta{}^\alpha & \Leftrightarrow \; \boldsymbol{T} = -\boldsymbol{T}^T \\
\textit{is antisymmetric then follows} & \\
I_{11} = 0, \;\; I_{12} = -I_{13} \;\; \textit{and} \;\; I_{14} = -I_{15} = 0. &
\end{array}
\tag{4.4.3}
$$

4.5 Special Tensors

Next, coming back to Section 2.2 some specific matrices are regarded to examine their tensor properties.

At first, with (4.2.2) the covariant metric $g_{\alpha\beta}$ and with (4.2.4) the contravariant Metric $g^{\alpha\beta}$ are considered. As ***metric tensors of the 2nd order***, their components are transformed with (4.4.2) from the Cartesian to the skewed reference system as follows:

$$
\begin{array}{ll}
g_{\alpha\beta} = g^i{}_\alpha * g^k{}_\beta * \delta_{ik} & g_{\alpha\beta} = g^i{}_\alpha * g^k{}_\beta * \mathbf{e}_i * \mathbf{e}_k \\
\textit{with } det(g_{\alpha\beta}) = g & \textit{according to } (4.2.3) \\
 & \\
g^{\alpha\beta} = g_i{}^\alpha * g_k{}^\beta * \delta^{ik} & g^{\alpha\beta} = g_i{}^\alpha * g_k{}^\beta * \mathbf{e}^i * \mathbf{e}^k \\
\textit{with } det(g^{\alpha\beta}) = 1/g \textit{ according to } (4.2.5) &
\end{array}
\;. \tag{4.5.1}
$$

According to (4.2.7), the ***mixed-variant metrics*** $g_\alpha{}^\beta$ and $g^\alpha{}_\beta$ ***do not exhibit tensor behavior*** following the transformation rules, and they are only useful as mixed δ-matrices for exchange of indices, as follows:

$$
\begin{array}{ll}
g_\alpha{}^\beta = g^i{}_\alpha * g_k{}^\beta * \delta_i{}^k = g^i{}_\alpha * g_i{}^\beta = \delta_\alpha{}^\beta & \delta_\alpha{}^\beta = g^i{}_\alpha * g_k{}^\beta * \mathbf{e}_i * \mathbf{e}^k \\
 & \\
g^\alpha{}_\beta = g_i{}^\alpha * g^k{}_\beta * \delta^i{}_k = g_i{}^\alpha * g^i{}_\beta = \delta^\alpha{}_\beta & \delta^\alpha{}_\beta = g_i{}^\alpha * g^k{}_\beta * \mathbf{e}^i * \mathbf{e}_k
\end{array}
\;. \tag{4.5.2}
$$

Next, the e-matrix (2.2.4) can be examined for its tensor behavior when transformed into skewed reference systems. Using the transformation rules (4.4.2), Eq. (2.2.5) can be modified as follows:

$$\varepsilon_{\alpha\beta\gamma} = e_{ikl} * g^i{}_\alpha * g^k{}_\beta * g^l{}_\gamma = \left| \delta^s{}_i \ \delta^s{}_k \ \delta^s{}_l \right| * g^i{}_\alpha * g^k{}_\beta * g^l{}_\gamma = \begin{vmatrix} g^0{}_\alpha & g^0{}_\beta & g^0{}_\gamma \\ g^1{}_\alpha & g^1{}_\beta & g^1{}_\gamma \\ g^2{}_\alpha & g^2{}_\beta & g^2{}_\gamma \end{vmatrix} .$$
$$with\ N_s = N_i = N_k = N_l = 3 \tag{4.5.3}$$

Due to (4.2.3) with
$det(\varepsilon_{\alpha\beta\gamma}) = \ det(g^i{}_\alpha) = \sqrt{g}$ for even permutations of (0,1,2),
$det(\varepsilon_{\gamma\beta\alpha}) = -det(g^i{}_\alpha) = -\sqrt{g}$ for odd permutations of (2,1,0),
$det(\varepsilon_{\alpha\beta\gamma}) = \ 0$ otherwise,
this transformation, compared to (2.2.4), provides the definitions of the ***covariant Levi-Civita tensor*** resp. ***ε-tensor***

$$\varepsilon_{\alpha\beta\gamma} = \sqrt{g} * e_{\alpha\beta\gamma} = \begin{cases} \sqrt{g}, \textit{for } (\alpha,\beta,\gamma) \textit{ as even permutations of } (0,1,2..), \\ -\sqrt{g}, \textit{for } (\alpha,\beta,\gamma) \textit{ as odd permutations of } (2,1,0..), \\ 0, \textit{ if at least two indices are the same} \end{cases} , \tag{4.5.4}$$

and the ***contravariant Levi-Civita tensor*** resp. ***ε-tensor***

$$\varepsilon^{\alpha\beta\gamma} = 1/\sqrt{g} * e^{\alpha\beta\gamma} = \begin{cases} 1/\sqrt{g}, \textit{for } (\alpha,\beta,\gamma) \textit{ as even permutations of } (0,1,2..), \\ -1/\sqrt{g}, \textit{for } (\alpha,\beta,\gamma) \textit{ as odd permutations of } (2,1,0..), \\ 0, \quad \textit{if at least two indices are the same} \end{cases} . \tag{4.5.5}$$

Now following Section 2.2, the tensor notation can be applied also to the ***cross product*** (2.2.7)

$$\mathrm{v}_\alpha = \varepsilon_{\alpha\beta\gamma} * b^\beta * c^\gamma \qquad \boldsymbol{v} = \boldsymbol{b} \times \boldsymbol{c} , \tag{4.5.6}$$

as well as to the ***scalar triple product*** according to (2.2.8)

$$\mathrm{s} = \varepsilon_{\alpha\beta\gamma} * a^\alpha * b^\beta * c^\gamma \qquad s = \boldsymbol{a} * (\boldsymbol{b} \times \boldsymbol{c}) . \tag{4.5.7}$$

The calculation of ***determinants*** can be achieved by use of invariants for ***(2x2)-tensors***, according to (2.2.10)

$$\det(\boldsymbol{A}) = 1/2 * \epsilon^{\alpha\beta} * \epsilon^{\gamma\delta} * A_{\alpha\gamma} * A_{\beta\delta} = 1/2 * \epsilon_{\alpha\beta} * \epsilon_{\gamma\delta} * A^{\alpha\gamma} * A^{\beta\delta} \tag{4.5.8}$$

and for ***(3x3)-tensors***, according to (2.2.11)

$$\boxed{det(A) = 1/6 * \epsilon^{\alpha\beta\gamma} * \epsilon^{\kappa\lambda\mu} * A_{\alpha\kappa} * A_{\beta\lambda} * A_{\gamma\mu} = 1/6 * \epsilon_{\alpha\beta\gamma} * \epsilon_{\kappa\lambda\mu} * A^{\alpha\kappa} * A^{\beta\lambda} * A^{\gamma\mu}} . \tag{4.5.9}$$

For the determination of various ε-identities and also the contravariant tensor basis, the ***Laplace expansion theorem*** (2.2.6) is crucial, which in tensor notation can be derived from (4.5.3), and forms the ***Kronecker delta*** resp. ***δ-matrix of 6th order***

$$\begin{aligned} &in\ \mathbb{R}^3:\ \delta_{\alpha\beta\gamma}{}^{\kappa\lambda\mu} = \varepsilon_{\alpha\beta\gamma} * \varepsilon^{\kappa\lambda\mu} = \begin{vmatrix} g^s{}_\alpha \\ g^s{}_\beta \\ g^s{}_\gamma \end{vmatrix} * \begin{vmatrix} g_s{}^\kappa & g_s{}^\lambda & g_s{}^\mu \end{vmatrix} \quad with\ N_S = 3 \\ &= \begin{vmatrix} \delta_\alpha{}^\kappa & \delta_\alpha{}^\lambda & \delta_\alpha{}^\mu \\ \delta_\beta{}^\kappa & \delta_\beta{}^\lambda & \delta_\beta{}^\mu \\ \delta_\gamma{}^\kappa & \delta_\gamma{}^\lambda & \delta_\gamma{}^\mu \end{vmatrix} = \delta_\alpha{}^\mu * \begin{vmatrix} \delta_\beta{}^\kappa & \delta_\beta{}^\lambda \\ \delta_\gamma{}^\kappa & \delta_\gamma{}^\lambda \end{vmatrix} - \delta_\beta{}^\mu * \begin{vmatrix} \delta_\alpha{}^\kappa & \delta_\alpha{}^\lambda \\ \delta_\gamma{}^\kappa & \delta_\gamma{}^\lambda \end{vmatrix} + \delta_\gamma{}^\mu * \begin{vmatrix} \delta_\alpha{}^\kappa & \delta_\alpha{}^\lambda \\ \delta_\beta{}^\kappa & \delta_\beta{}^\lambda \end{vmatrix} \end{aligned} . \tag{4.5.10}$$

By contraction the ***Grassmann ε-identity***, also refered to as the ***contracted ε-identity***, arises

$$\begin{aligned} &in\ \mathbb{R}^3:\ \delta_{\alpha\beta}{}^{\kappa\lambda} = \varepsilon_{\alpha\beta\gamma} * \varepsilon^{\kappa\lambda\gamma} = \begin{vmatrix} \delta_\alpha{}^\kappa & \delta_\alpha{}^\lambda \\ \delta_\beta{}^\kappa & \delta_\beta{}^\lambda \end{vmatrix} = \delta_\alpha{}^\kappa * \delta_\beta{}^\lambda - \delta_\alpha{}^\lambda * \delta_\beta{}^\kappa \\ &\qquad respectively \\ &in\ \mathbb{R}^2:\ \delta_{\alpha\beta}{}^{\kappa\lambda} = \varepsilon_{\alpha\beta} * \varepsilon^{\kappa\lambda} = \delta_\alpha{}^\kappa * \delta_\beta{}^\lambda - \delta_\alpha{}^\lambda * \delta_\beta{}^\kappa \end{aligned} \tag{4.5.11}$$

as ***Kronecker delta*** resp. ***δ-matrix of 4th order***.

This relationship is i.a. useful in deriving further identities for the cross product $\mathbf{c} = \mathbf{a} \times \mathbf{b} \quad \Leftrightarrow \quad c^\mu = \varepsilon^{\kappa\lambda\mu} * a_\kappa * b_\lambda$:

$$\begin{aligned} &as\ the\ \textbf{Lagrange identity}\ (\mathbf{a}\times\mathbf{b})\cdot(\mathbf{c}\times\mathbf{d}) = (\mathbf{a}\cdot\mathbf{c})(\mathbf{b}\cdot\mathbf{d}) - (\mathbf{b}\cdot\mathbf{c})(\mathbf{a}\cdot\mathbf{d}) \quad \Leftrightarrow \\ &\quad \boldsymbol{\varepsilon}_{\alpha\beta\mu} * a^\alpha * b^\beta * \boldsymbol{\varepsilon}^{\kappa\lambda\mu} * c_\kappa * d_\lambda = a^\alpha * b^\beta * \boldsymbol{\delta}_{\alpha\beta}{}^{\kappa\lambda} * c_\kappa * d_\lambda \\ &\quad = a^\alpha * c_\alpha * b^\beta * d_\beta - b^\beta * c_\beta * a^\alpha * d_\alpha , \\ \\ &or\ the\ \textbf{Grassmann identity}\ \mathbf{c}\times(\mathbf{a}\times\mathbf{b}) = (\mathbf{b}\cdot\mathbf{c})\mathbf{a} - (\mathbf{a}\cdot\mathbf{c})\mathbf{b} \quad \Leftrightarrow \\ &\quad c^\beta * \boldsymbol{\varepsilon}_{\alpha\beta\mu} * \boldsymbol{\varepsilon}^{\kappa\lambda\mu} * a_\kappa * b_\lambda = c^\beta * \boldsymbol{\delta}_{\alpha\beta}{}^{\kappa\lambda} * a_\kappa * b_\lambda \\ &\quad = b_\beta * c^\beta * a_\alpha - a_\beta * c^\beta * b_\alpha , \\ \\ &and\ the\ \textbf{Jacobi identity}\ \mathbf{c}\times(\mathbf{a}\times\mathbf{b}) + \mathbf{a}\times(\mathbf{b}\times\mathbf{c}) + \mathbf{b}\times(\mathbf{c}\times\mathbf{a}) \equiv \mathbf{0} \quad \Leftrightarrow \\ &\quad \boldsymbol{\varepsilon}_{\alpha\beta\mu} * (c^\beta * a_\kappa * b_\lambda + a^\beta * b_\kappa * c_\lambda + b^\beta * c_\kappa * a_\lambda) * \boldsymbol{\varepsilon}^{\kappa\lambda\mu} \\ &\quad = (c^\beta * a_\kappa * b_\lambda + a^\beta * b_\kappa * c_\lambda + b^\beta * c_\kappa * a_\lambda) * \boldsymbol{\delta}_{\alpha\beta}{}^{\kappa\lambda} \\ &\quad = a_\alpha * (c^\beta * b_\beta - b^\beta * c_\beta) + b_\alpha * (a^\beta * c_\beta - c^\beta * a_\beta) \\ &\quad + c_\alpha * (b^\beta * a_\beta - a^\beta * b_\beta) \equiv 0 \end{aligned} . \tag{4.5.12}$$

Further contractions of the Grassmann ε-identity (4.5.11), considering (4.5.4) and (4.5.5), lead finally to the following Kronecker delta as ***δ-identities of 2nd*** and ***0th order***

$$\boxed{\begin{array}{ll} 2*\delta_{\alpha}{}^{\beta}=\varepsilon_{\alpha\lambda\gamma}*\varepsilon^{\beta\lambda\gamma}=e_{\alpha\lambda\gamma}*e^{\beta\lambda\gamma} & \delta_{\alpha}{}^{\beta}=\varepsilon_{\alpha\lambda}*\varepsilon^{\beta\lambda}=e_{\alpha\lambda}*e^{\beta\lambda} \\ 2*\delta_{\alpha}{}^{\alpha}=\varepsilon_{\alpha\beta\gamma}*\varepsilon^{\alpha\beta\gamma}=e_{\alpha\beta\gamma}*e^{\alpha\beta\gamma}=6 & \delta_{\alpha}{}^{\alpha}=\varepsilon_{\alpha\beta}*\varepsilon^{\alpha\beta}=e_{\alpha\beta}*e^{\alpha\beta}=2 \end{array}}\ . \tag{4.5.13}$$

These identities have the important property that both the contravariant basis vectors and the contravariant metric can be determined directly from the covariant basis — solely through the rotation of the basis axes — without the need for matrix inversions.

For this purpose with (4.5.13), the ***two-dimensional δε-identity***[1] can be prepared as follows:

$$\begin{array}{l} \delta^{ik}=e^{i}{}_{l}*e^{kl}=e^{ir}*\delta_{rl}*e^{kl} \\ respectively \\ \delta_{i}{}^{k}=e_{il}*e^{kl}=e_{ir}*\delta^{r}{}_{l}*e^{kl}\ . \end{array}$$

The transformation (4.4.2) into the skewed coordinate system leads then, with index renaming, to the ***two-dimensional Tensorbasis Transformation***[2] with

$$\boxed{\begin{array}{ll} the\ \mathbf{metric\ tensor} & g^{\alpha\beta}=\varepsilon^{\alpha\gamma}*g_{\gamma\lambda}*\varepsilon^{\beta\lambda} \\ and & \\ the\ \mathbf{basis\ vectors} & g_{i}{}^{\beta}=e_{ir}*g^{r}{}_{\gamma}*\varepsilon^{\beta\gamma} \end{array}}\ , \tag{4.5.14}$$

where the contravariant ε-tensor may be computed in advance following (4.5.5).

In the lower equation, a rotation of the covariant basis vectors $g^{r}{}_{\gamma}$ by 90° is performed by multiplying with the left-hand e-matrix e_{ir}. Subsequently, the right-hand ε-tensor $\varepsilon^{\beta\gamma}$ corrects the orientation of the contravariant basis vectors and normalizes their length by a factor of $1/\sqrt{g}$. A similar approach applies to the upper equation.

With (4.5.13), the ***three-dimensional δε-identity***[1] can be applied in tensor notation as follows:

$$\begin{array}{l} 2*\delta^{ik}=e^{i}{}_{lm}*e^{klm}=e^{irp}*\delta_{rl}*\delta_{pm}*e^{klm} \\ respectively \\ 2*\delta_{i}{}^{k}=e_{ilm}*e^{klm}=e_{irp}*\delta^{r}{}_{l}*\delta^{p}{}_{m}*e^{klm}\ . \end{array}$$

The transformation (4.4.2) into the skewed coordinate system leads then, with index renaming, to the ***three-dimensional Tensorbasis Transformation***[2] with

$$\boxed{\begin{array}{ll} the\ \mathbf{metric\ tensor} & g^{\alpha\beta}=1/2*\varepsilon^{\alpha\gamma\kappa}*g_{\gamma\lambda}*g_{\kappa\mu}*\varepsilon^{\beta\lambda\mu} \\ and & \\ the\ \mathbf{basis\ vectors} & g_{i}{}^{\beta}=1/2*e_{irp}*g^{r}{}_{\gamma}*g^{p}{}_{\kappa}*\varepsilon^{\beta\gamma\kappa} \end{array}}\ . \tag{4.5.15}$$

4.6 Object-oriented Basis Vectors

After explaining the fundamentals of skew-angled tensor bases and their metrics in Section 4.2, the question now arises of how these quantities can be efficiently determined for specific numerical applications. For this purpose, the properties of covariant basis vectors must be captured, and their metric tensors need to be calculated. This eventually yields contravariant basis vectors with their metric. The associated calculation algorithm can be organized in a general form in a new tensor class BASIS, utilizing the matrix classes from Chapter 3.

[1] As a preparation for appropriate tensor transformations, the new $\delta\varepsilon$-identities are another modification of the Grassmann ε-identity.
[2] ***These new tensor basis transformations are an alternative approach to evaluate the contravariant basis vectors and their metric tensor by means of the Levi-Civita tensor.***

4.6.1 Tensor Class BASIS

The class BASIS [Meissner 2023] is based on the two matrix classes MATRIX and VEKTOR and relies on their methods (member functions).

With the four constructors of class BASIS, one-, two-, and three-dimensional basis vectors with elements of type 'double' can be created. Input and output are handled through the overloaded operators '>>' and '<<' .

Example:

```
VEKTOR V0(3),V1(3);                              // build covariant vectors
V0(0)=5.0;    V0(1)=0.0;    V0(2)=2.0;  // predefine with values
V1(0)=5.0;    V1(1)=5.0;    V1(2)=2.0;
//
BASIS &g2 = *new BASIS(V0,V1);      // Constructor: create instance g2
//
MATRIX &cov_basis = g2.covariant_basis();              // covariant basis
cout<<"covariant␣basis␣vectors␣"<<cov_basis;
//
MATRIX metric(2,2);
g2.covariant_metric(metric);                           // covariant metric
cout<<"covariant␣metric␣"<<metric;
//
double volume = g2.sqrt_g();                           // volume element
cout<<"volume=␣"<<volume<<'\n';
//
delete &g2;                   // Destructor: delete BASIS-instance g2
delete &cov_basis;            //             delete MATRIX cov_basis
```

Declaration file (header file): Basis.h

Listing 4.1: Class BASIS

```
// ================ Tensor Class BASIS ================================
// Declaration file (Header file)
// --------------------------------
class BASIS
{
// ====== Declaration of Prototypes ==================================
// Tensor class for basis vectors, metrics,
// volume elements, and epsilon tensors
// Basis vectors g(i,alpha) and matrices mat(i,j,k,l) are
// stored column-wise !!!
// Indexing of alpha,i,j,k,l starts at 0 !!!
// cf. U. Breymann, C++, Carl Hanser Verlag 1993, p.186 ff
// ===================================================================
// Copyright: Univ.-Prof. Dr.-Ing. Udo F. Meissner, Erkrath/Darmstadt
// Advanced teaching and study program - Commercial use not permitted.
// The author assumes no liability for incompleteness and errors!
// Version:  2023
// Compiler: C++
// ===================================================================
```

```
//                    Public Member Functions
//                    =======================
   public:
// =======
//
// Constructor_0 -> BASIS g(mat)
// ==============================
// mat: Matrix with covariant basis vectors
// ----------------------------------------
   BASIS(MATRIX &mat);
//
// Constructor_1 -> BASIS g(vek1)
// ===============================
// vek1: Vector with covariant basis vector 1
// ------------------------------------------
   BASIS(VEKTOR &vek1);
//
// Constructor_2 -> BASIS g(vek1,vek2)
// ====================================
// vek1: Vector with covariant basis vector 1
// vek2: Vector with covariant basis vector 2
// ------------------------------------------
   BASIS(VEKTOR &vek1,VEKTOR &vek2);
//
// Constructor_3 -> BASIS g(vek1,vek2,vek3)
// ========================================
// vek1: Vector with covariant basis vector 1
// vek2: Vector with covariant basis vector 2
// vek3: Vector with covariant basis vector 3
// ------------------------------------------
   BASIS(VEKTOR &vek1,VEKTOR &vek2,VEKTOR &vek3);
//
// Destructor -> delete &g
// =======================
   ~BASIS();
//
// Covariant basis vectors -> g.covariant_basis()
// ==============================================
   MATRIX &kovariante_basis();
   MATRIX &covariant_basis ();
//
// Contravariant basis vectors -> g.contravariant_basis()
// ======================================================
   MATRIX &kontravariante_basis();
   MATRIX &contravariant_basis ();
//
// Length or area or volume element -> g.sqrt_g()
// ==============================================
// sqrt_g = sqrt(Determinant of g(alpha,beta))
// -------------------------------------------
   double &wurzel_g();
   double &sqrt_g  ();
//
// Covariant metric -> g.covariant_metric(mat)
```

```
// =============================================
// g(alpha,beta) = g(i,alpha) * g(i,beta)
// Result: Determinant (g)
// ---------------------------------------
   double kovariante_metrik(MATRIX &mat);
   double covariant_metric (MATRIX &mat);
//
// Contravariant metric -> g.contravariant_metric(mat)
// =====================================================
// g(alpha,beta) = g(i,alpha) * g(i,beta)
// ---------------------------------------
   void kontravariante_metrik(MATRIX &mat);
   void contravariant_metric (MATRIX &mat);
//
// Covariant epsilon-tensor -> g.covariant_epsilon_tensor(mat)
// ===========================================================
   void kovarianter_epsilon_tensor(MATRIX &mat);
   void covariant_epsilon_tensor  (MATRIX &mat);
//
// Contravariant epsilon-tensor-> g.contravariant_epsilon_tensor(mat)
// ==================================================================
   void kontravarianter_epsilon_tensor(MATRIX &mat);
   void contravariant_epsilon_tensor  (MATRIX &mat);
};
// ==================================================================
```

4.6.2 Object-oriented Function METRIC[1]

This section includes the calculation function METRIC [Meissner 2023], to illustrate algorithms for determining basis vectors in Euclidean space and their metrics through specific examples. For this purpose, the tensor class BASIS explained in Section 4.6.1 is used.

The following quantities will be determined:
- covariant basis vectors and their covariant metric,
- contravariant basis vectors and their contravariant metric,
- volume, area, resp. line elements.

The individual program steps correspond in their sequence to the theoretical explanations presented in Section 4.2, and can therefore be followed step by step with the inserted comments.
The handling of its output stream outFILE by the associated main program FEM_Demo [Meissner 2023] is outlined in Section 5.4.

Published below is the object-oriented program code of the function METRIC in the C++ programming language for the

Metric of Basis Vectors with Perspective Graphics, cf. Chapter 5:

Listing 4.2: Function METRIC: Tensor/Matrix Algorithms for Metric Evaluation

```
// =================================================================
// Function METRIC for Demonstrating the Tensor Class BASIS
//          ------
// C++ - MATRIX Objects and Methods for Tensor/Matrix Calculus
// =================================================================
// Copyright: Univ.-Prof. Dr.-Ing. Udo F. Meissner, Erkrath/Darmstadt
// Advanced teaching and study program - Commercial use not permitted.
// The author assumes no liability for incompleteness and errors!
// Version:  2023
// Compiler: C++
// =================================================================
// FEM: Determine BASIS Vectors and Metrics:
// -----------------------------------------
// Co- and contravariant basis vectors, metric- and epsilon-tensors
// =============== Include Files ====================================
#include "pch.h"                                // pre-compiled header file
// ------------
// #include "BASES_3D.h"
            // 1/2/3-dimensional Basis und Metric        BASES_3D
// #include "Basis.h"
            // 1/2/3-dimensional Basis and Metric        BASIS
// #include "Matrix.h"
            // 4-dimensional double Rectangle Matrices   MATRIX
// #include "Vektor.h"
```

[1] Function METRIC provides some comments about original member function calls used in C++ source code of classes available on the WWW [Meissner 2023].

```
             // 1-dimensional Vectors                       VEKTOR
// =================================================================
//  METRIK
int METRIC(ofstream& outFILE, ifstream& inFILE, int branch)
{                          // outFILE logs the data input
                           // inFILE provides the file input
                           // branch= 0 end of METRIC without BASES_3D
                           //        > 0 continue with BASES_3D
    // -------------------------------------------------------------
    // ofstream outFILE;
    // string output = "Output-File";
    // outFILE.open(output, ios::app);
    // cout << "Output file: " << output << endl << endl;
    // =============================================================
    bool ERROR;
    ERROR = false;
    //*** ERROR = true;
    if (ERROR) std::cout << "Error␣=␣" << ERROR << endl;
    // --------------------------------------------------
    int dim_i = 1, dim_alpha = 1;
        bool read_in = false, g1Yes = true, g2Yes = true;
    //*** bool read_in = true;
    //
    // Dimension dim_i of Euclidean space
    //           to read basis vectors mat = g[dim_i, dim_alpha]
    // #########################################################
    std::cout << "Dimension␣of␣Cartesian␣coordinates:" << endl;
    std::cout << "dim_i␣=␣1␣one-dimensional␣" << endl;
    std::cout << "␣␣␣␣␣␣=␣2␣two-dimensional␣" << endl;
    std::cout << "␣␣␣␣␣␣=␣3␣three-dimensional␣" << endl;
    std::cout << "Enter␣dim_i:␣";
    std::cin >> dim_i;                      outFile << dim_i << endl;
    std::cout << endl;
    // -------------------------------------------------------------
    //
    MATRIX mat(3, 3);
    // -------------------------------------------------------------
    // input
    std::cout
        << "Enter␣basis␣vectors␣individually?" << endl;
    std::cout << "->␣1␣=␣yes␣/␣0␣=␣no␣(default):␣";
    std::cin >> read_in;                 outFile << read_in << endl;
    // -------------------------------------------------------------
    // switch (dim_i)
    // ===============
    if (dim_i == 1)
    {   // one-dimensional Euclidean space
        // ================================
        std::cout << "One-dimensional␣Euclidean␣Space" << endl;
        std::cout << "================================" << endl;
        // Number of Basis Vectors
        dim_alpha = 1;
        //
        // mat.dimension_aendern(dim_i, dim_alpha);
```

```
        mat.change_dimension (dim_i, dim_alpha);
     // ----------------------------------------
     // Pre-set
     // =======
        VEKTOR g0(dim_i);
        g0 = 5.0;
        // ----------------------------------------------------------
        if (read_in != 0)
        { // Read Basis Vector g0
          // ###################
            //
            // Read Vector
            std::cout << dim_i <<
            "␣Components␣of␣basis␣vector␣g0:␣"<<endl<<endl;
            std::cout << "Vector␣g0:␣" << endl;
            std::cin >> g0;                    outFile << g0 << endl;
        }
        // Store Basis Vectors in mat = g[dim_i, dim_alpha]
        // ----------------------------------------------------------
        // mat.dimension_aendern(dim_i, dim_alpha);
           mat.change_dimension (dim_i, dim_alpha);
        // ----------------------------------------
           mat = g0;
  }
  else
  {
     if (dim_i == 2)
     { // two-dimensional Euclidean space
       // ===============================
         std::cout << "Two-dimensional␣Euclidean␣Space" <<endl;
         std::cout << "===============================" <<endl;
         // Number of Basis Vectors
         dim_alpha = 2;
         // ------------------------
         // Pre-set, see Parallelogram Element
         // ==================================
            VEKTOR g0(dim_i), g1(dim_i);
            g0(0) = 2.5;     g1(0) = 0.0;
            g0(1) = 0.0;     g1(1) = 5.0;
        // ----------------------------------------------------------
        if (read_in != 0)
        { // Read Basis Vectors Individually
         std::cout << "Read␣basis␣vectors␣g0,␣g1␣individually"
                   <<"␣as␣linearly␣independent␣" << endl << endl;
               // Read Basis Vector g0
               // ###################
               std::cout << dim_i
                  << "␣Components␣of␣basis␣vector␣g0:␣"
                  << endl << endl;
               std::cout << "Vector␣g0:␣" << endl;
              std::cin >> g0;                    outFile << g0 << endl;
               dim_alpha = 1;
            // ----------------------------------------------------------
        std::cout << "Read␣second␣basis␣vector␣g1?" << endl;
```

```
    std::cout << "->␣1␣=␣yes␣/␣0␣=␣no␣(default):␣"
                                                            << endl;
        std::cin >> g1Yes;             outFile << g1Yes << endl;
        // ---------------------------------------------------------
            if (g1Yes)
            {   // Read Basis Vectors g1
                // ####################
             std::cout << dim_i
               << "␣Components␣of␣basis␣vector␣g1:␣"
               << endl << endl;
                std::cout << "Vector␣g1:␣" << endl;
                std::cin >> g1;           outFile << g1 << endl;
                dim_alpha = 2;
            }
        }
    else {std::cout <<
        "Pre-set,␣see␣Parallelogram␣Element␣"<< endl; };

    // Store Basis Vectors in mat = g[dim_i, dim_alpha]
    // -------------------------------------------------
    // mat.dimension_aendern(dim_i, dim_alpha);
       mat.change_dimension (dim_i, dim_alpha);
    // -----------------------------------------
       mat = g0;
       //  if (g1Yes) {mat.teilmatrix_hinein(0, 1) = g1;}
           if (g1Yes) {mat.partialmatrix_in (0, 1) = g1;}
}
else
{ // three-dimensional vector space
  // ==============================
  std::cout << "Three-dimensional␣Euclidean␣Space" << endl;
  std::cout << "=================================" << endl;
    // Number of basis vectors
    dim_alpha = 3;
    //
    // Read Basis Vector mat(dim_i, dim_i)
    // ##################################
    //
    // Pre-set, see Triangular Element
    // ==============================
    VEKTOR g0(dim_i), g1(dim_i), g2(dim_i);
       g0(0) = 6.0;     g1(0) =  1.0;    g2(0) = 1.0;
       g0(1) = 2.0;     g1(1) = 12.0;    g2(1) = 2.0;
       g0(2) = 1.0;     g1(2) =  1.0;    g2(2) = 1.0;
    // ---------------------------------------------------------
       if (read_in != 0)
       {   // Read Basis Vectors Individually
      std::cout<<"Read␣basis␣vectors␣g0,␣g1,␣g2␣individually"
               <<"␣as␣linearly␣independent␣" << endl << endl;
           // Read Basis Vector g0
           // ####################
     std::cout
      << dim_i <<"␣Components␣of␣Basis␣Vector␣g0:␣"
               << endl << endl;
```

```
        std::cout << "Vector␣g0:␣" << endl;
        std::cin >> g0;               outFile << g0 << endl;
        dim_alpha = 1;
        // ---------------------------------------------
        std::cout << "Read␣second␣basis␣vector␣g1?"  << endl;
        std::cout << "->␣1␣=␣yes␣/␣0␣=␣no␣(default):␣"<<endl;
        std::cin >> g1Yes;           outFile << g1Yes << endl;
        // ---------------------------------------------
           if (g1Yes)
           {
            // Read Basis Vector g1
            // ####################
              std::cout<< dim_i
                   << "␣Components␣of␣basis␣vector␣g1:␣"
                   << endl << endl;
              std::cout << "Vector␣g1:␣" << endl;
              std::cin >> g1;             outFile << g1 << endl;
              //*** dim_alpha = dim_alpha +1;
           };
           // ------------------------------------------
       dim_alpha = dim_alpha + 1;
       //
       std::cout << "Read␣third␣basis␣vector␣g2?"
                 << endl;
       std::cout
        << "->␣1␣=␣yes␣/␣0␣=␣no␣(default):␣" << endl;
       std::cin >> g2Yes;            outFile << g2Yes << endl;
           // ------------------------------------------
       if (g2Yes)
       {
              // Read Basis Vector g2
              // ####################
      std::cout << dim_i
               <<"␣Components␣of␣basis␣vector␣g2:␣"
               << endl << endl;
           std::cout << "Vector␣g2:␣" << endl;
           std::cin >> g2;                outFile << g2 << endl;
           dim_alpha = dim_alpha + 1;
       }
       // -----------------------------------------------
       }
       else {std:: cout
              << "Pre-set,␣see␣Triangle␣Element␣"<<endl;}
       //
   // Store Basis Vectors in mat = g[dim_i, dim_alpha]
   // -----------------------------------------------------
   // mat.dimension_aendern(dim_i, dim_alpha);
      mat.change_dimension (dim_i, dim_alpha);
   // ------------------------------------
      mat = g0;
      // if (g1Yes)  mat.teilmatrix_hinein(0, 1) = g1;
         if (g1Yes)  mat.partialmatrix_in (0, 1) = g1;
      // if (g2Yes)  mat.teilmatrix_hinein(0, 2) = g2;
         if (g2Yes)  mat.partialmatrix_in (0, 2) = g2;
```

```
      };
    };
    // -----------------------------------------------------------------
    std::cout << endl;
    std::cout<<"Matrix␣of␣Basis␣Vectors␣g[dim_i,␣dim_alpha]␣=␣"
              << mat <<endl;
                                                  outFile << "" << endl;
    // =================================================================
    // Create Basis
    // ============
    BASIS& basis = *new BASIS(mat);
    //
    // Covariant Basis Vectors
    // -----------------------
    MATRIX& cov_basis = *new MATRIX(dim_i, dim_alpha);
    // cov_basis = basis.kovariante_basis();
       cov_basis = basis.covariant_basis ();
    std::cout << "Covariant␣Basis␣Vectors␣" << cov_basis;
                                                  outFile << "" << endl;
    //
    // sqrt(g)
    // ---------
    // double sqrt_g = basis.wurzel_g();
       double sqrt_g = basis.sqrt_g  ();
    std::cout << endl;
    std::cout << "Square␣Root(g)␣=␣" << sqrt_g << endl << endl;
    //
// Contravariant Basis Vectors
// ----------------------------
    MATRIX& contra_basis = *new MATRIX(dim_i, dim_alpha);
    // contra_basis = basis.kontravariante_basis();
       contra_basis = basis.contravariant_basis ();
    std::cout << "Contravariant␣Basis␣Vectors␣" << contra_basis;
                                                  outFile << "" << endl;
    //
    // Test -> Identity Matrix
    // -----------------------
    MATRIX& iden = *new MATRIX(dim_alpha, dim_alpha);
    //
        iden = cov_basis.dimension(1, dim_i, dim_alpha) *
                         contra_basis.dimension(1, dim_i, dim_alpha);
    std::cout << endl;
    std::cout << "Test:␣Identity␣Matrix␣=␣covariant␣basis␣*␣"<<
                                      "contravariant␣basis␣" << iden;
                                                  outFile << "" << endl;
    //
    // Covariant Metric
    // ----------------
    MATRIX& cov_metric = *new MATRIX(dim_alpha, dim_alpha);
    //
    // double g_value = basis.kovariante_metrik(cov_metric);
       double g_value = basis.covariant_metric (cov_metric);
    std::cout << endl;
    std::cout << "Covariant␣Metric␣" << cov_metric;
```

```
                                                      outFile << "" << endl;
    //
    std::cout << endl;
    std::cout << "Determinant(g)␣=␣" << g_value << endl<< endl;
    //
    // Contravariant Metric
    // ---------------------
    MATRIX& contra_metric = *new MATRIX(dim_alpha, dim_alpha);
    //
    // basis.kontravariante_metrik(contra_metric);
       basis.contravariant_metric (contra_metric);
    std::cout << "Contravariant␣Metric␣" << contra_metric;
                                                      outFile << "" << endl;
    //
    // Test -> Identity Matrix
    // -----------------------
    iden = cov_metric.dimension(1, dim_alpha, dim_alpha) *
                contra_metric.dimension(1, dim_alpha, dim_alpha);
    std::cout << endl;
    std::cout << "Test:␣Identity␣Matrix␣=␣covariant␣metric␣*"
                                 <<"␣contravariant␣metric␣" << iden;
                                                      outFile << "" << endl;
    // ===================================================================
    std::cout << endl;
    std::cout << "End␣METRIC" << endl;
    std::cout << "==========" << endl;
    //
    std::cout << "PAUSE:␣";
    // ####################
    char c;
    std::cin >> c;                                    outFile << c << endl;
    // ===================================================================
    // Plot Basis Vektors perspectively
    if (branch > 0)
    {// BASIS_3D
       BASES_3D(outFile, inFILE, cov_basis, sqrt_g, contra_basis);
       return 2;
    };
    // ===================================================================
    return 1;
};
// ======================================================================
```

As a demonstration example for the utilization of the Tensor class BASIS, three independent vectors are chosen in the function METRIC, which span a tetrahedron in three-dimensional Euclidean space, as later illustrated in Section 7.2 using the example of the triangular element (cf. Fig. 7.2). Covariant basis vectors *cov_basis* and the corresponding metric *cov_metric*, as well as contravariant basis vectors *contra_basis* with their metric *contra_metric* and the volume element *sqrt_g* are determined.

Listing 4.3: Demonstration Example METRIC of Tensor Class BASIS

```
METRIC with Basis Vectors
=========================
```

```
Numerical Determination of Basis Vectors, Metrics and Volume Elements
=====================================================================
Dimension of Cartesian coordinates:
dim_i = 1 one-dimensional
      = 2 two-dimensional
      = 3 three-dimensional
Enter dim_i: 3

Enter basis vectors individually?
-> 1 = yes / 0 = no (default): 0
Three-dimensional Euclidean Space
=================================
Pre-set, see Triangle Element

Matrix of Basis Vectors g[dim_i, dim_alpha] =
[ (3*3) * (1*1) ]-MATRIX:
Columns 0 to 2
 6.000e+00  1.000e+00  1.000e+00
 2.000e+00  1.200e+01  2.000e+00
 1.000e+00  1.000e+00  1.000e+00

Pause -> enter a character:

Covariant Basis Vectors
[ (3*3) * (1*1) ]-MATRIX:
Columns 0 to 2
 6.000e+00  1.000e+00  1.000e+00
 2.000e+00  1.200e+01  2.000e+00
 1.000e+00  1.000e+00  1.000e+00

Pause -> enter a character:

Square Root(g) = 50

Contravariant Basis Vectors
[ (3*3) * (1*1) ]-MATRIX:
Columns 0 to 2
 2.000e-01  5.551e-17 -2.000e-01
-2.220e-16  1.000e-01 -1.000e-01
-2.000e-01 -2.000e-01  1.400e+00

Pause -> enter a character:

Test: Identity Matrix = covariant basis * contravariant basis
[ (3*3) * (1*1) ]-MATRIX:
Columns 0 to 2
 1.000e+00  2.498e-16 -7.994e-15
-2.109e-15  1.000e+00  8.882e-16
 1.110e-16 -2.776e-17  1.000e+00

Pause -> enter a character:

Covariant Metric
```

```
[ (3*3) * (1*1) ]-MATRIX:
Columns 0 to 2
 4.100e+01  3.100e+01  1.100e+01
 3.100e+01  1.460e+02  2.600e+01
 1.100e+01  2.600e+01  6.000e+00

Pause -> enter a character:

Determinant(g) = 2500

Contravariant Metric
[ (3*3) * (1*1) ]-MATRIX:
Columns 0 to 2
 8.000e-02  4.000e-02 -3.200e-01
 4.000e-02  5.000e-02 -2.900e-01
-3.200e-01 -2.900e-01  2.010e+00

Pause -> enter a character:

Test: Identity Matrix = covariant metric * contravariant metric
[ (3*3) * (1*1) ]-MATRIX:
Columns 0 to 2
 1.000e+00 -8.882e-16 -3.553e-15
-7.105e-15  1.000e+00  2.132e-14
-4.441e-16 -4.441e-16  1.000e+00

Pause -> enter a character:

End METRIC
==========
```

References

de Boer/Schroeder 2012. de Boer, R.; Schröder, J.: Tensor Calculus for Engineers: Analytical and Computational Aspects. Springer, Berlin (2012)

Duschek/Hochrainer 1968. Duschek, A.; Hochrainer, A.: Grundzüge der Tensorrechnung in analytischer Darstellung. Springer, Berlin (1968)

Green/Zerna 1954. Green, A. E.; Zerna, W.: Theoretical Elasticity. At The Clarendon Press, Oxford (1954)

Klingbeil 1966. Klingbeil, E.: Tensorrechnung für Ingenieure. Bibliographisches Institut, Mannheim (1966)

Internet Links

Meissner 2023. Meißner, U. F.: Object-oriented Tensor and Matrix Classes (2023)
`https://rebrand.ly/Matrix_Classes`

Chapter 5
Three-dimensional Visualization of Vector Objects

Abstract

With the mathematical fundamentals of tensor calculus and its object-oriented implementation in numerical algorithms described in the preceding chapters, the objects in three-dimensional Euclidean space have been captured. They can be managed accordingly and represented through transformations in different views. The illustrations shown clarified these relationships in the spatial context.
For more complex situations, there is generally the challenge of developing a spatial understanding of how the objects are arranged and how they relate to each other. This requires appropriate geometric models that can be spatially examined from different sights to understand the positioning of objects, detect errors, and exclude unauthorized collisions.
For this purpose parallel and central perspectives on technical objects are of common use in engineering. Therefore in the following sections, such a concept is converted into object-oriented methods and implemented in the adequate classes PERSPEKTIVE and PyPLOT with elementary graphic tools based on the universal programming language Python being available in Visual Studio [Microsoft 2023]. Its application is demonstrated with the graphics function BASES_3D. Figure 5.5 displays the intended clarity in a typical stereo vision.

In the field of mechanics and engineering, the visualization of complex technical objects is of main importance, because the mathematically-numerical models serve to represent real structures, such as buildings, machines, flow regimes etc., with their physical properties. These models are used to dimension their components, prepare technical designs and solutions for real projects, and analyze the behavior of structures a priori through mathematical-numerical simulations.

Accordingly, it is necessary to visualize the treated objects in three-dimensional space. Understanding the complexity of the relationships should be sharpened and trained in this way. In the present case, this task is limited to the captured point, line, and tensor objects, so that a representation by so-called wire frame models seems appropriate to spatially depict composite structures.

For visualization of these objects, the commonly used parallel and central perspectives in engineering are suitable. With the present tensor and matrix calculus, this goal is addressed and achieved by object-oriented software methods in the following sections. Figure 5.5 displays the intended clarity in a typical stereo vision.

U. F. Meissner, *Tensor Calculus with Object-Oriented Matrices for Numerical Methods in Mechanics and Engineering*, https://doi.org/10.1007/978-3-031-59302-4_5

5.1 Central and Parallel Perspectives in Three-Dimensional Space

In the Euclidean vector space $\mathbb{R}^3$ of the world coordinate system $\mathbf{e}_i$, the perspective on spatial point, line, and vector objects is achieved by mapping these structures onto a two-dimensional projection surface or screen. This principle is illustrated in Fig. 5.1.

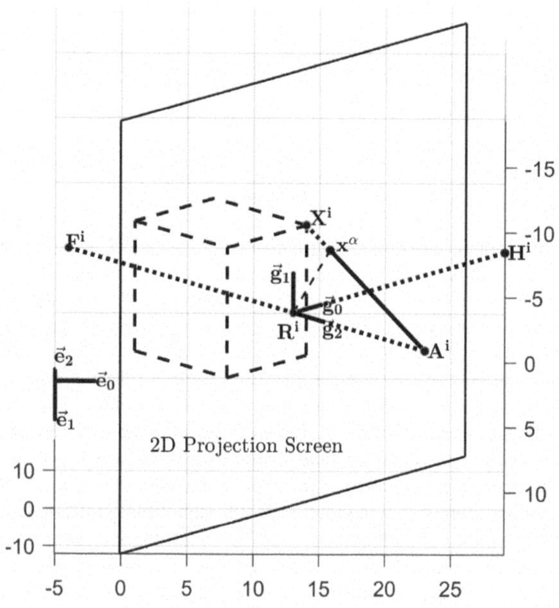

Fig. 5.1: Central perspective - Mapping of the 3D object onto the 2D projection screen

From the eye point A^i, the line of sight is directed towards the vanishing point F^i and passes through the projection screen at the reference point R^i, defined as the origin of the 2D image vector space $\mathbb{R}^2$ with basis vectors $\mathbf{g}_\alpha$. The horizon point H^i determines the horizontal alignment of the image and simultaneously establishes the eye-to-screen spacing because the horizon is perpendicular to the line of sight $A^i - F^i$.

Looking from the viewpoint A^i to the object point X^i, its mapping as the intersection point x^i on the projection screen is obtained. The basic task is to use tensor calculus to determine the world coordinates of the image point x^i with respect to the reference point R^i and then transform them into the image basis $\mathbf{g}_\alpha$. The image of the object, represented here as a cube, can be drawn with the image coordinates x^α on the projection screen as a wireframe. This basic graphics are sufficiently informative for the spatial illustration of the elementary point, line, and vector objects discussed here, revealing the arrangement of objects in space and their mutual association succinctly. It also allows for viewing the objects from different perspectives by changing the viewing direction.

With this central perspective, it is also possible to simultaneously view the objects with two eyes using the same principle. This variation creates a stereo image on the projection screen in two colors (e.g. cyan and red), which, when viewed with a corresponding red/cyan-colored glasses, produces an image that floats in space and vividly reflects the distances between object parts. Such a stereo perspective is implemented in Section 5.4 using object-oriented methods to visually strengthen the understanding of tensor calculus objects.

The ordinary parallel perspective is derived from the central perspective by placing the vanishing point fictitiously into the infinite distance, so that all rays run parallel in direction of the projection screen. Both forms of representation can be easily selected, solely based on the manner by which the image points are transformed.

5.1.1 Determination of the Image Basis

The determination of the Cartesian image basis $\mathbf{g}_\alpha$ is illustrated in Fig. 5.2 and is related to the coordinate origin O^i based on the specifications for the eye, vanishing, and horizon points.

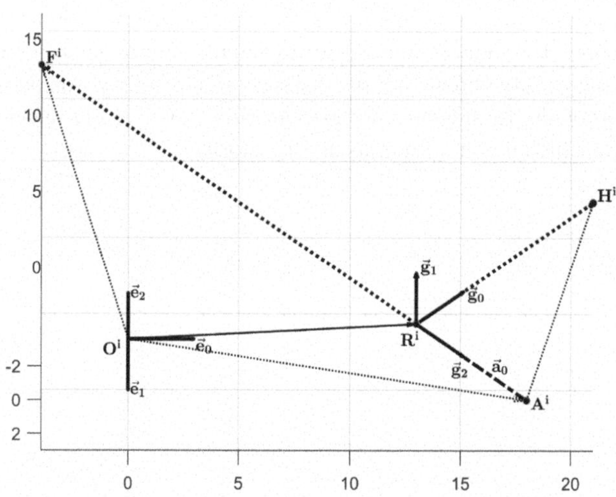

Fig. 5.2: Projection screen - Determination of the image basis vectors

First, the ***basis vector*** $\mathbf{g}_2$ is formed, which is necessary to measure spacings perpendicular to the projection screen.

$$\boxed{g^i{}_2 = (A^i - F^i)/|A^k - F^k| \quad \Leftrightarrow \quad \mathbf{g}_2 = (\mathbf{A}-\mathbf{F})/|\mathbf{A}-\mathbf{F}|} \tag{5.1.1}$$

With this, the ***eye spacing*** a_0 from the viewer's eye to the projection screen

$$\boxed{a_0 = g^i{}_2 * (A^i - H^i) \quad \Leftrightarrow \quad a_0 = \mathbf{g}_2 * (\mathbf{A}-\mathbf{H})} \tag{5.1.2}$$

and the spacing vector $\mathbf{a}_0$

$$a^i{}_0 = -a_0 * g^i{}_2$$

can be determined.

As a result, the position vector to the ***reference point*** R^i of the image basis is given by:

$$\boxed{R^i = A^i + a^i{}_0 \quad \Leftrightarrow \quad \mathbf{R} = \mathbf{A} + \mathbf{a}_0}\,. \tag{5.1.3}$$

Finally, the two ***basis vectors*** $\mathbf{g}_0$ and $\mathbf{g}_1$ of the projection screen $\mathbb{R}^2$ can be calculated

$$\boxed{\begin{array}{lll} g^i{}_0 = (H^i - R^i)/|H^k - R^k| & \Leftrightarrow & \mathbf{g}_0 = (\mathbf{H} - \mathbf{R})/|\mathbf{H} - \mathbf{R}| \\ \textit{and} & & \\ g^i{}_1 = \varepsilon^i{}_{kl} * g^k{}_2 * g^l{}_0 & \Leftrightarrow & \mathbf{g}_1 = \mathbf{g}_2 \times \mathbf{g}_0 \end{array}} \; . \qquad (5.1.4)$$

5.1.2 Determination of the Image Points

First, the image points for the parallel perspective are determined. Fig. 5.3 shows the projection of the object point X^i onto the projection screen with the image point x^i_p.

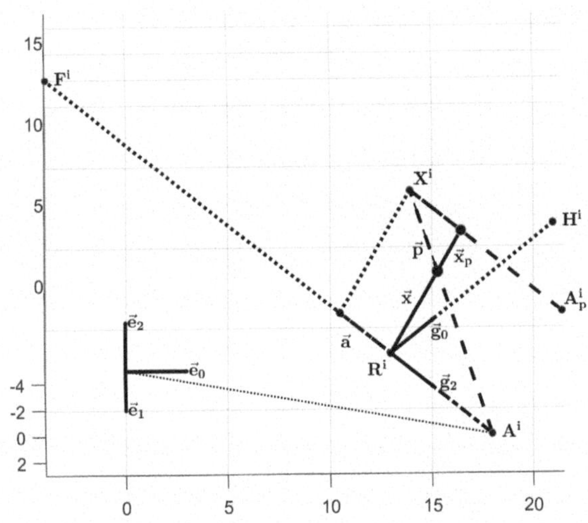

Fig. 5.3: Perspective - Determination of the image points

From the eye point A^i, the line of sight **p** is directed towards the object point X^i

$$p^i = X^i - A^i$$

and the ***distance*** p from the eye of the observer to the object point is given by

$$\boxed{\mathrm{p} = |X^i - A^i| \quad \Leftrightarrow \quad p = |\mathbf{X} - \mathbf{A}|}, \tag{5.1.5}$$

while the ***object spacing*** a from the eye of the observer to the object point is calculated by

$$\boxed{\mathrm{a} = -\, \mathrm{g}^i{}_2 * p^i \quad \Leftrightarrow \quad a = -\mathbf{g}_2 * \mathbf{p}}. \tag{5.1.6}$$

With the vector **a**

$$a^i = -a * g^i{}_2$$

the ***image vector*** $\mathbf{x}_p$ - measured in world coordinates - is given by

$$\boxed{\mathrm{x}^i_p = p^i - a^i \quad \Leftrightarrow \quad \mathbf{x}_p = \mathbf{p} - \mathbf{a}}. \tag{5.1.7}$$

For the parallel perspective, the vector $\mathbf{x}_p$ is mapped onto the projection screen in its true length.

For the central perspective, however, the vector $\mathbf{x}_p$, according to Fig. 5.3, is to be shortened to the ***image vector*** $\mathbf{x}$ on the projection screen. This shortening results from the ratio of the eye spacings a_0 and a using equations 5.1.2 and 5.1.6.

$$\alpha = a_0/a \quad \textit{for the } \textbf{\textit{central perspective}}$$
$$\alpha = 1 \quad \textit{for the } \textbf{\textit{parallel perspective}} \qquad (5.1.8)$$
$$x^i = \alpha * x^i_p \quad \Leftrightarrow \quad \mathbf{x} = \alpha * \mathbf{x_p}$$

The components x^i must now be transformed with the contravariant basis vectors from 5.1.4 into the reference system of the image basis $\mathbf{g}_\alpha$ to obtain the ***image point*** x^α on the projection screen.

$$x^\alpha = x^i * g_i{}^\alpha$$
$$\textit{for } \alpha \in \mathbb{N}\,[0,1] \textit{ and } i \in \mathbb{N}\,[0,1,2] \qquad (5.1.9)$$

The image points x^α can now be presented using 2D graphics in the image reference system of the projection screen, as explained in Section 5.4, supplemented with line and vector objects.

To guide the observer to the optimal position for viewing the central perspective, it is helpful to specify the spacing a_0 from the eye point A^i to the reference point R^i below as an optimal measure.

5.1.3 Stereo Projection for Dual Eyes

For stereo projection, the transformations from Section 5.1.2 need to be performed twice - separately for the left eye (red lens) and the right eye (cyan lens) - with a human eye separation distance of approximately 6.75 cm. This results in double images on the projection screen in cyan and red colors, as shown in Figure 5.4.

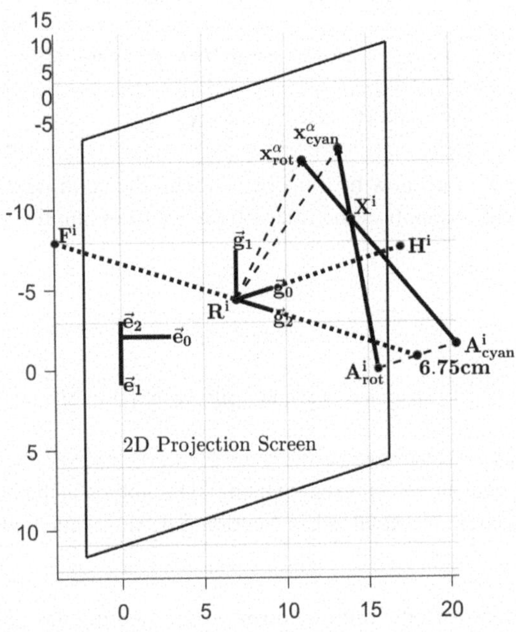

Fig. 5.4: Perspective - Stereo projection in red and cyan colors

With the red/cyan-colored glasses, the left eye observes only the cyan color, and the right eye only the red color. In this way, an impressive stereo vision is achieved for a comprehensive assessment of the depicted objects, while viewing the image with the optimal eye spacing a_0 from the projection screen.

5.2 Tensor Class PERSPEKTIVE

The tensor class PERSPEKTIVE[1] implements the mathematical fundamentals described in Section 5.1 into object-oriented methods for generating parallel and central perspectives. Within the constructor, it is necessary to store the position of the eyes and the horizon in Euclidean space and capture the properties of the projection relative to the projection screen. The method *transformation* performs the transformation of three-dimensional world coordinates to two-dimensional image coordinates. To normalize the image size, it determines also the maximum and minimum image coordinates. Additionally, the spacings and distances from the object points to the eye point are stored in order to determine which object points are in front of or behind the projection screen.

The class PERSPEKTIVE [Meissner 2023] is based on the two matrix classes MATRIX and VEKTOR and uses their object-oriented methods.
The constructor determines whether to create a parallel perspective or a central perspective with the options 'Mono' or 'Stereo.' In the 'Stereo' option, the projection is made onto a screen that has to be placed behind the object points. The image coordinates obtained as a result of the transformation are of type 'double'. The input and output of text and numbers are done using the inherited operators '>>' and '<<'.

Example:

```
// Eye point A                                  // World coordinates of A
double a[3] = {8, -35, 15}; VEKTOR A(a, 3);      // pre-set C array a
// Vanishing point F                            // World coordinates of F
double f[3] = {5,   5,  5}; VEKTOR F(f, 3);      // pre-set C array f
// Horizon point H                              // World coordinates of H
double h[3] = {15, 15,  5}; VEKTOR H(h, 3);      // pre-set C array h

int mode = 1;                                    // Central Perspective
int num_eyes = 2;                                // Stereo Projection
//
// Constructor
PERSPEKTIVE *Persp = new PERSPEKTIVE(A, F, H, mode, num_eyes);
// Determining the image coordinates a of the eye point A
VEKTOR a_image(2);
// Determining the eye spacing a0 to the zero-reference point R
double a0_eye_spacing;
a_image = Persp->eye(a0_eye_spacing);
// Destructor
delete Persp;    // Delete the instance Persp using the destructor '~'
```

Declaration file (header file): Perspekt.h

Listing 5.1: Class PERSPEKTIVE

```
// ==================================================================
// Tensor Class PERSPEKTIVE for 3D Central and Parallel Perspective
// ==================================================================
```

[1] engl. PERSPECTIVE

```
// Declaration file (header file)
class PERSPEKTIVE
{
// ==== Declaration of Prototypes ====================================
// Matrices are stored column-wise!!!
// Indexing of mk starts at 0!!!
// see U. Breymann: C++. Carl Hanser Verlag (1993), p.186 ff
// ==================================================================
// Copyright: Univ.-Prof. Dr.-Ing. Udo F. Meissner, Erkrath/Darmstadt
// Advanced teaching and study program - Commercial use not permitted.
// The author assumes no liability for incompleteness and errors!
// Version:  2023
// Compiler: C++
// ==================================================================
//                        Public member functions
//                        =======================
   public:
// =======
//
// Constructor -> PERSPEKTIVE per(A, F, H, mode, eyes);
// ================================================
// A(3): x-y-z world coordinates of the eye point
// F(3): x-y-z world coordinates of the vanishing point
// H(3): x-y-z world coordinates of the horizon point
// mode: 1=Central perspective / 0=Parallel perspective
// eyes (for central perspective): 1=Mono / 2=Stereo
// ------------------------------------------------------------------
// The view is from the eye point A to the vanishing point F.
// From the horizon point H, the horizon line
//                                      is perpendicular to the line A-F.
// The intersection point R forms the zero-reference point
//                        of the image coordinates on the projection screen,
// determines its position, and gives the spacing to the eye.
// For 'Stereo', the projection screen should be behind the objects!
// ------------------------------------------------------------------
   PERSPEKTIVE(VEKTOR& A, VEKTOR& F, VEKTOR& H,int art=1,int eyes=1);
//
// Destructor -> delete &per;
// ==========================
// Check the position of the projection plane (Horizon H)
//                                                   behind the object points
   ~PERSPEKTIVE();
//
// Eye point -> a_image=per.eye(&a0_spacing_R);
// ================================================
// Result -> a_image(2):   1-2 Image coordinates of the eye point
// Result -> a0_spacing_R: Spacing of the eye from the zero-
//                                  reference point of the projection screen
// ------------------------------------------------------------------
   MATRIX& auge(double& a0_spacing_R);
   MATRIX& eye (double& a0_spacing_R);
//
// Horizon point -> h_image=per.horizon();
// ==========================================
```

```
// Result -> h_image(2): 1-2 Image coordinates of the horizon point
// ------------------------------------------------------------------
   MATRIX& horizont();
   MATRIX& horizon ();
//
// Image point -> x_image=per.transformation(X_world,
//                                               spac_dist_X, points);
// ===================================================================
// X_world(3, points): x-y-z world coordinates of the object points
// Result -> x_image(2, eyes, points): 1-2 Coordinates
//                                               of the image points
// Result -> spac_dist_X(points [, points]): Spacing [, Distance]
//                                 of the object points X from the eye
// points: Number of object points
// -------------------------------------------------------------------
   MATRIX& transformation(MATRIX& X, MATRIX& spac_dist_X,
                                                   int points = 1);
//
// Image window coordinates -> per.window(min1, min2, max1, max2);
// ===================================================================
// Result -> min1, min2: Minimum of all 1-2 image coordinates
// Result -> max1, max2: Maximum of all 1-2 image coordinates
// -------------------------------------------------------------------
   void fenster(double& fmin1, double& fmin2, double& fmax1,
                                                 double& fmax2);
   void window (double& fmin1, double& fmin2, double& fmax1,
                                                 double& fmax2);
};
// ===================================================================
```

5.3 Graphics Class PyPLOT

Building upon the previously described methods of perspective representation, the treated 3D objects are transformed into the image coordinates of the projection screen. Based on this procedure, the class PyPLOT [Meissner 2023] completes the two-dimensional graphical representation of these objects on the projection screen.

For this purpose, the C++ library ***matplotlib-cpp*** [Microsoft 2023] is chosen, which can be implemented in Visual Studio as a packaged interface module (wrapper). It efficiently displays the present point, line, and vector objects in black/white or in color based on the *Python* programming language. As minimally necessary, here only the *point, line, and tripod* methods are implemented for drawing. The constructor scales the images in height and width. The input and output of text and numbers are done using the inherited operators '>>' and '<<'.

An application example follows in the next Section (5.4) and is elaborately commented there.

Declaration file (header file): PyPLOT.h

Listing 5.2: Class PyPLOT

```
// ==================================================================
// Graphic Class PyPLOT by Graphics Python/C++ Wrapper matplotlibcpp
// ==================================================================
// -> Notes on installation in Visual Studio:
// https://stackoverflow.com/questions/66507618/
//                     running-python-matplotlibcpp-in-visual-studio-2019
// -> Documentation notes:
// Python Graphics: https://matplotlib-cpp.readthedocs.io/en/latest/
// -> Usage notes:
// matplotlibcpp: https://github.com/lava/matplotlib-cpp
// ==================================================================
// Declaration file (header file)
// --------------------------------
class PyPLOT
{
// ====== Declaration of prototypes ==================================
// Matrices are stored column-wise!!!
// Indexing of mk starts at 0!!!
// see U. Breymann: C++. Carl Hanser Verlag (1993), p.186 ff
// ==================================================================
// Copyright: Univ.-Prof. Dr.-Ing. Udo F. Meissner, Erkrath/Darmstadt
// Advanced teaching and study program - Commercial use not permitted.
// The author assumes no liability for incompleteness and errors!
// Version:  2023
// Compiler: C++
// ==================================================================
//              Public member functions
//              =======================
   public:
```

```
// =======
//
// Constructor -> PyPLOT py(factor0, factor1, circle);
// ====================================================
// fact0 / fact1 -> Scaling factors
//                                  for the image window coordinates x / y
// circle -> Radius for possible drawing. Circumference of image
//             points resp. for the distance of the text label in [mm]
// ---------------------------------------------------------------------
   PyPLOT(double fact0 = 1.0,double fact1 = 1.0,double circle = 1.0);
//
// Destructor -> delete& py;
// =========================
   ~PyPLOT();
// ---------------------------------------------------------------------
// Print Image Title
// =================
// text  -> Image title
// ---------------------
   void bild_titel (string text);
   void image_title(string text);
//
// Show Image
// ==========
   void zeige_bild();
   void show_image();
//
// Draw Point -> py.point
// ======================
// A_point      -> Image coordinates of point A
//                                       with A(dimension, eyes, [points])
// eye          -> 0 / 1 left / right eye
// [point]      -> 0 resp. (0 / 1 / 2) -> Image point
//
// color        -> Color of the point representation
// marker       -> Symbol of the point
//                                             (according to Python graphics)
// label        -> Text label of the point
// number       -> Numerical value for the point (e.g., eye spacing)
// ---------------------------------------------------------------------
// MATRIX A_point is two-dimensional and contains only one point
   void punkt(MATRIX& A_punkt,
        int auge, string farbe, string marker, string bezeichner,
                                                 double zahl   = 0.0);
   void point(MATRIX& A_point,
        int eye,  string color, string marker, string label,
                                                 double number = 0.0);
// ----------------
// MATRIX A_point is three-dimensional and contains multiple
//                                          points in the third dimension
   void punkt(MATRIX& A_punkt, int punkt_a,
        int auge, string farbe, string marker,string bezeichner,
                                                 double zahl   = 0.0);
   void point(MATRIX& A_point, int point_a,
```

```
        int eye, string color, string marker, string label,
                                                  double number = 0.0);
// ===================================================================
//
// Draw Line -> py.line
// ====================
// A_point     -> Image coordinates of starting point A
//                                 with A(dimension, eyes, [points])
// [point_a]   -> 0 resp. (0 / 1 / 2) -> Image point a
// E-point     -> Image coordinates of end point E
//                                 with E(dimension, eyes, [points])
// [point_e]   -> 0 resp. (0 / 1 / 2) -> Image point e
// eye         -> 0 / 1 left / right eye
// color       -> Color of the line representation
// linestyle   -> Line style
//                                   (according to Python graphics)
// label       -> Text label of the end point E
// -------------------------------------------------------------------
// MATRIX A_point, E_point are three-dimensional and
//                 contain multiple points in the third dimension
   void linie(MATRIX& A_vek, int punkt_a, MATRIX& E_vek, int punkt_e,
              int auge, string farbe, string lstil, string bezeichner);
   void line (MATRIX& A_vek, int point_a, MATRIX& E_vek, int point_e,
              int eye, string color, string linestyle, string label);
// ----------------
// MATRIX A_point, E_point are two-dimensional with only one point
   void linie(MATRIX& A_vek,                 MATRIX& E_vek,
              int auge, string farbe, string lstil, string bezeichner);
   void line (MATRIX& A_vek,                 MATRIX& E_vek,
              int eye, string color, string linestyle, string label);
// ===================================================================
//
// Draw Tripod -> num_vek = py.tripod
// ==================================
// A_point -> Image coordinates of the starting point
//                                   A (dimension, eyes)
// E_point -> Image coordinates of the end points
//                                   E (dimension, eyes, vectors)
// eye         -> 0 / 1 left / right eye
// color       -> Color of the vector representation
// linestyle   -> Line style of the vector representation
//                                   (according to Python graphics)
// label   -> Text label "E", "G" or "g" of the end points E;
//            will be supplemented by Cartesian, covariant or
//                                          contravariant indices
// Result -> Number of tripod vectors (1-/2-/3-dimensional)
// point   -> Point of the matrix vector vek
// -------------------------------------------------------------------
   int dreibein(MATRIX& A_punkt, MATRIX& E_punkte, int auge,
                      string farbe, string lstil, string bezeichner);
   int tripod  (MATRIX& A_point, MATRIX& E_points, int eye,
                      string color, string linestyle, string label);
};
// ===================================================================
```

5.4 Graphics Function BASES_3D[1] for Basis Vectors and Tensor Objects

In this section, the graphics function BASES_3D is introduced to highlight the advantages of perspective visualization using one-, two-, or three-dimensional basis vectors with the previously explained classes PERSPEKTIVE and PyPLOT. For the implementation of the C++ program code, the methods of tensor/matrix calculus are consistently employed. This results in an object-oriented structure that is clear and well-organized.

With the two beginning arguments of the function BASES_3D, the handling of its input/output streams is pre-organized by the associated main program FEM_Demo [Meissner 2023]. In the first step of the application, inFILE reads the input data from the user's terminal, and outFILE backs up the read data into the log file outFILE, while the produced results of the function are displayed by the standard output std::cout. In a second run of the application, the saved input data may be read from the log file by changing its name to inFILE, and the function results are then stored into outFILE, as displayed in Listing 5.4.

By the second part of the argument list, the results of the function METRIC from Section (4.6.2) are transferred into the function BASES_3D. These are

- the matrix *cov_basis* of covariant basis vectors,
- the line/area/volume element *sqrt_g*,
- the matrix *contra_basis* of contravariant basis vectors.

In the first part of the function, the transformations of 3D world coordinates to 2D image coordinates of the projection screen are performed using the methods of the PERSPEKTIVE class. This is done in the following steps:

- Select the type of perspective (central or parallel perspective),
- Define the spatial positions of eye, vanishing, and horizon points,
- Shift the basis vectors for clarity towards the center of the image using a position vector,
- Set the number of eyes (mono or stereo display),
- Perform an optical stretching of Cartesian and contravariant basis vectors by $\sqrt{g}$ to keep these vectors visible at the image in comparison to generally longer covariant basis vectors,
- Perform the transformation of all world coordinates into image coordinates.

In the second part of the function, the graphical representation of objects on the projection screen is then performed using the methods of class PyPLOT. The following steps are necessary:

- Normalize the image to minimum/maximum window coordinates,

[1] Function BASES_3D provides some comments for original member function calls used in C++ source code of classes available on the WWW [Meissner 2023].

- Map the objects using the methods *point, line, tripod* in different foreground colors:
- For parallel and central perspective (mono with one eye) in black,
- For central perspective (stereo with two eyes) in cyan (left eye) and red (right eye).

In the following passage, the object-oriented program code of the function BASES_3D in the C++ program language is listed for the

Three-dimensional Perspective of Basis Vectors:

Listing 5.3: Function BASES_3D: Tensor/Matrix Algorithms of Central and Parallel Perspectives

```
// ===========================================================================
// Function BASES_3D
//          for Central and Parallel Perspectives of Basis Vectors
//          ----------------------------------
// C++ - Matrix Objects and Methods for Tensor/Matrix Calculus
// ===========================================================================
// Copyright: Univ.-Prof. Dr.-Ing. Udo F. Meissner, Erkrath/Darmstadt
// Advanced teaching and study program - Commercial use not permitted.
// The author assumes no liability for incompleteness and errors!
// Version:  2023
// Compiler: C++
// ===========================================================================
// FEM: Perspective on 1-/2-/3-dimensional basis vectors:
// ----------------------------------------------------------
// Cartesian reference system and co- and contravariant basis vectors
// =============== Include Files =============================================
#include "pch.h"                              // pre-compiled header file
// -------------
// #include "Perspekt.h"
            // Perspective on graphic objects            PERSPEKTIVE
// #include "PyPLOT.h"
            // C++ wrapper for Python graphics           PyPLOT
// #include "Matrix.h"
            // 4-dimensional double-rectangle matrices   MATRIX
// #include "Vektor.h"
            // 1-dimensional vectors                      VEKTOR
// ===========================================================================
int BASES_3D(ofstream& outFILE, ifstream& inFILE,
             MATRIX& cov_basis, double sqrt_g, MATRIX& contra_basis)
{                      // outFILE logs the data input
                       // inFILE provides the file input
    // Input parameter:
                      // cov_basis    covariant basis vectors
                      // sqrt_g       line- / area- / volume element
                      // contra_basis contravariant basis vectors
    // -----------------------------------------------------------------------
    // ofstream outFILE;
    // string Output = "Output File";
    // outFILE.open(Output, ios::app);
    // cout << "Output File: " << Output << endl << endl;
    // =======================================================================
    bool ERROR;
```

```
    ERROR = false;
    //*** ERROR = true;
    if (ERROR) std::cout << "Error␣=␣" << ERROR << endl;
    // -----------------------------------------------
    int dim_i = 1, dim_alpha = 1, dummy;
    bool input = false;
    //
//  Determine the dimension of the basis vectors
//  ---------------------------------------------
//  cov_basis.dimension_ermitteln(dim_i, dim_alpha, dummy, dummy);
    cov_basis.determine_dimension(dim_i, dim_alpha, dummy, dummy);
// ===================================================================
// Display basis vectors in Euclidian space by 3D-perspective
// ===================================================================
  std::cout << "Skip␣3D-perspective␣->␣exit␣program?␣" << endl;
  std::cout << "->␣1␣=␣yes␣/␣0␣=␣no␣(default):␣" << endl;
  std::cout << "===============================␣";
  std::cin >> input;                        outFILE << input << endl;
  // =================================================================
  int z_persp;
  if (input == 0)
  { // =================== Block: Python Graphics ==================
    // Set graphics mode -> initialize the Python instance
    //*** std::cout << "Py_Initialize" << endl;
    //
       Py_Initialize();
    // ================
      int type_persp, eyes, num_eyes, num_points;
      // Number of vector points = Dimension alpha of basis vectors
      num_points = dim_alpha;
      // Default for vector lengths
      double LENGTH = 10.0;
      // #################
      // New Perspective
      VEKTOR A3(3), F3(3), H3(3), O3(3);
     // ===============================
  do
  { // ============== Block: New Perspective ===============
    // Type of perspective
       std::cout << endl << "Enter␣type␣of␣perspective" << endl;
    std::cout <<
    "1␣=␣Central␣perspective␣/␣0␣=␣Parallel␣perspective␣/␣-1␣=␣end:␣"
             << endl;
    std::cout <<
    "============================================================␣";
       std::cin >> type_persp;          outFILE << type_persp << endl;
    // ==============================================================
    // Enter eye point, vanishing point, and point on the horizon
    // ============================================================
       std::cout <<
  "Perspective:␣The␣eye␣A␣is␣directed␣towards␣the␣vanishing␣point␣F,"
                 << endl;
       std::cout <<
            "the␣horizon␣point␣H␣determines␣the␣horizontal␣line␣and,"
```

```
                << endl;
        std::cout <<
  "spans␣by␣SPACING␣the␣projection␣screen,␣perpendicular␣to␣A␣-␣F␣!"
                << endl;
      std::cout << endl << "Enter␣eye␣point␣A:␣" << endl;
   //                    ==================
     // Default
      A3(0) = LENGTH * 0.5 +  3.0;  A3(1) = LENGTH * 0.5 - 50.0;
      A3(2) = LENGTH * 0.5 + 10.0;
      std::cout << "Default␣coordinates␣for␣A:␣" << A3 << endl;
                                                  outFILE << "" << endl;
      std::cout << "Change?␣->␣1␣=␣yes␣/␣0␣=␣no␣(default):␣"<< endl;
      std::cout << "=======================================␣";
      std::cin >> input;                        outFILE << input << endl;
      if (input == 1)
      {
        std::cout << "Enter␣x0,␣x1,␣x2:" << endl;
        std::cin >> A3;                           outFILE << A3 << endl;
      }
      if (ERROR) std::cout << "Coordinates␣A:␣" << A3 << endl;
   // ----------------------------------------------------------------
      std::cout << endl <<
          "Enter␣vanishing␣point␣F␣(in␣center␣of␣object):␣" << endl;
   //       ======================
     // Default
     F3(0)= LENGTH * 0.5; F3(1)= LENGTH * 0.5; F3(2)= LENGTH * 0.5;
     std::cout << "Default␣coordinates␣F:␣" << F3 << endl;
                                                  outFILE << "" << endl;
     std::cout << "Change?␣->␣1␣=␣yes␣/␣0␣=␣no␣(default):␣" << endl;
     std::cout << "=======================================␣";
     std::cin >> input;                          outFILE << input << endl;
     if (input == 1)
     {
       std::cout << "Enter␣x0,␣x1,␣x2:" << endl;
       std::cin >> F3;                            outFILE << F3 << endl;
     };
     if (ERROR) std::cout << "Coordinates␣F:␣" << F3 << endl;
     // --------------------------------------------------------------
     std::cout << endl <<
         "Enter␣horizon␣point␣H␣(determines␣the␣horizon):␣" << endl;
     // Default
     H3(0)= LENGTH + 5.0; H3(1)= LENGTH + 6.0; H3(2)= LENGTH * 0.5;
     std::cout << "Default␣coordinates␣H:␣" << H3 << endl;
                                                  outFILE << "" << endl;
     std::cout << "Change?␣->␣1␣=␣yes␣/␣0␣=␣no␣(default):␣" << endl;
     std::cout << "=======================================␣";
     std::cin >> input;                          outFILE << input << endl;
     if (input == 1)
     {
       std::cout << "Enter␣x0,␣x1,␣x2:" << endl;
       std::cin >> H3;                            outFILE << H3 << endl;
     }
     if (ERROR) std::cout << "Coordinates␣H:␣" << H3 << endl;
     // --------------------------------------------------------------
```

```
// Input position vector for improved representation of basis vectors
// ====================================================================
     std::cout<<
  "For␣a␣better␣overview,␣" << endl;
     std::cout<<
  "␣␣␣␣␣␣␣␣␣␣␣␣the␣basis␣vectors␣are␣shifted␣by␣the␣position␣vector"
           << endl;
     std::cout<<
  "-------------------------------------------------------------------"
           << endl << endl;
     //
     std::cout << "Enter␣endpoint␣of␣position␣vector␣O␣" << endl;
     //
    O3 = 0.0;
    O3(0)= LENGTH * 0.25; O3(1)= LENGTH * 0.25; O3(2)= LENGTH * 0.25;
     std::cout << "Default␣coordinates␣O:␣" << O3 << endl;
                                         outFILE << "" << endl;
     std::cout << "Change?␣->␣1␣=␣yes␣/␣0␣=␣no␣(default):␣" << endl;
     std::cout << "======================================␣";
     std::cin >> input;                    outFILE << input << endl;
     if (input == 1)
     {
       std::cout << "Enter␣x0,␣x1,␣x2:" << endl;
       std::cin >> O3;                      outFILE << O3 << endl;
     };
     if (ERROR) std::cout << "Coordinates␣O:␣" << O3 << endl;
     //
     // ===============================================================
// Transformation of basis vectors into 2D-space of projection screen
// ====================================================================
     while (type_persp >= 0)      // Programm-Ende
     { //========== Block: Number of Eyes =======================
     // Enter number of eyes
     // ====================
        num_eyes = 1;
     if (type_persp >= 1)
     {
     std::cout << endl <<
           "Enter␣number␣of␣eyes:␣1␣=␣Mono␣/␣2␣=␣Stereo:␣" << endl;
     std::cout <<
           "============================================␣";
     std::cin >> num_eyes;              outFILE << num_eyes << endl;
     if (num_eyes > 2) num_eyes = 2;
     if (num_eyes < 1) num_eyes = 1;
     };
     if (ERROR) std::cout << "Number␣of␣eyes:␣" << num_eyes << endl;
// --------------------------------------------------------------------
//
// Constructor call PERSP for transforming all coordinates
   PERSPEKTIVE& PERSP = *new PERSPEKTIVE(A3, F3, H3,
                                         type_persp, num_eyes);
// ====================================================================
//
// Image coordinates of eye point A with distance
```

```
//                                                  from the projection plane
// ========================================================================
   MATRIX a2(2);
   double a0_spacing_A;
   //
   // a2 = PERSP.auge(a0_spacing_A);
      a2 = PERSP.eye (a0_spacing_A);
   //
   std::cout << std::endl <<
"SPACING␣from␣the␣eye␣point␣A␣to␣the␣reference␣point␣0" << endl;
   std::cout <<
"␣␣␣␣␣␣␣␣␣␣␣␣␣␣␣␣␣␣␣␣␣␣␣␣␣␣␣␣␣␣of␣the␣projection␣screen:␣" << endl;
   std::cout << a0_spacing_A << std::endl;
                                                   outFILE << "" << endl;
//
if (ERROR) std::cout << "Image␣coordinates␣of␣the␣eye␣point:␣" << a2;
// ========================================================================
//
// Image coordinates of horizon point H
// ====================================
   VEKTOR h2(2);
   //
   // h2 = PERSP.horizont();
      h2 = PERSP.horizon ();
   //
   if (ERROR) std::cout << "Image␣coordinates␣of␣the␣horizon␣point:␣"
                       << h2;
  //
  // ======================================================================
  // Cartesian and contravariant basis are optically enlarged in the
  // graphics by sqrt_g as mean length of covariant basis vectors
  // ######################################################################
  //
     double            FACTOR = sqrt_g;
     if (dim_i == 2) FACTOR = sqrt(sqrt_g);
     if (dim_i == 3) FACTOR = pow(sqrt_g,   1.0/3.0);
  // ##################################################
   std::cout << std::endl;
   std::cout <<
   "Optical␣stretching␣of␣the␣contravariant␣and␣Cartesian␣bases"
             <<endl;
   std::cout << "by␣the␣FACTOR␣=␣" << FACTOR << endl << endl;
   std::cout <<"Change?␣->␣1␣=␣yes␣/␣0␣=␣no␣(default):␣"
             <<endl;
   std::cout << "=================================================␣";
   std::cin >> input;                  outFILE << input << std::endl;
   if (input == 1)
   {
    std::cout << "Enter␣FACTOR:" << endl;
    std::cin  >> FACTOR;                   outFILE << FACTOR << endl;
    input = 0;
   };
   // ---------------------------------------------------------------------
   //
```

```
    // Cartesian coordinate system
    // ----------------------------
    MATRIX X3(3, 3);
    X3(0, 0) = X3(1, 1) = X3(2, 2) = FACTOR;
    //
    // Image points of the Cartesian coordinate system
    // ================================================
    MATRIX x2(2, num_eyes, 3);
    MATRIX a_distance_X(3);
    //
    x2 = PERSP.transformation(X3, a_distance_X, 3);
    //
    if (ERROR) std::cout<<"Image␣coordinates␣of␣the␣Cartesian␣tripod:␣"
                        << x2;
    //
    // Image points of the position vector O
    // ======================================
// Initial point 0 = Origin U of Cartesian coordinate system
// =============
    VEKTOR U3(3);
    MATRIX u2(2, num_eyes);
    MATRIX a_distance_U(1);
    //
    u2 = PERSP.transformation(U3, a_distance_U);
    //
    if (ERROR) std::cout << "Image␣coordinates␣of␣the␣origin:␣"
                         << u2 << endl;
    // Endpoint = origin O of Basis vectors
    // ========
    //
    MATRIX o2(2, num_eyes);
    MATRIX a_distance_O(1);
    //
    if (ERROR) std::cout << "␣Coordinates␣of␣the␣position␣vector␣O␣="
                         << O3 << endl;
    //
    o2 = PERSP.transformation(O3, a_distance_O, 1);
    //
    if (ERROR) std::cout<<"Image␣coordinates␣of␣the␣position␣vector:␣"
                         << o2 << endl;
    // ===============================================================
    //
    // Image points of the covariant basis vectors
    // ============================================
    // all Basis vectors defined in perspectives are following as 3D
    //
       MATRIX G3(3,3);
    // store covariant basis vectors
    // G3.untermatrix_hinein(0, 0) = cov_basis;
       G3.submatrix_in      (0, 0) = cov_basis;
    //
       if (ERROR) std::cout << "Covariant␣3D␣basis␣=␣"
                            << G3 << endl;
    // ----------------------------------------------
```

```
// Shift endpoints by position vector
// G3.teilmatrix_hinein(0, 0) += O3;
   G3.partialmatrix_in (0, 0) += O3;
// G3.teilmatrix_hinein(0, 1) += O3;
   G3.partialmatrix_in (0, 1) += O3;
// G3.teilmatrix_hinein(0, 2) += O3;
   G3.partialmatrix_in (0, 2) += O3;
   //
   if (ERROR) std::cout << "Covariant␣3D-basis␣+␣O3␣=␣"
                 << G3 << std::endl;
   //
   MATRIX cov_g2(2, num_eyes, num_points);
   MATRIX a_distance_G(num_points);
   //
   cov_g2 = PERSP.transformation(G3, a_distance_G, num_points);
   //
   if (ERROR) std::cout <<
      "Image␣coordinates␣of␣the␣covariant␣basis␣vectors:␣"
                       << cov_g2 << std::endl;
   // ================================================================
   // *****************************
      ERROR = true;
   // *****************************

   // Image points of the contravariant basis vectors
   // ==================================================
   // Store-in contravariant basis vectors
   if (ERROR) std::cout << "contravariant␣basis␣=␣"
                       << contra_basis << endl;
   //
   G3 = 0.0;
   // G3.untermatrix_hinein(0, 0) = contra_basis * FACTOR;
      G3.submatrix_in       (0, 0) = contra_basis * FACTOR;
   //
   if (ERROR) std::cout << "Contravariant␣3D␣basis␣=␣"
                       << G3 << endl;
   // --------------------------------------------------
   // Shift endpoints by position vector
   // G3.teilmatrix_hinein(0, 0) += O3;
      G3.partialmatrix_in (0, 0) += O3;
   // G3.teilmatrix_hinein(0, 1) += O3;
      G3.partialmatrix_in (0, 1) += O3;
   // G3.teilmatrix_hinein(0, 2) += O3;
      G3.partialmatrix_in (0, 2) += O3;
   //
   if (ERROR) std::cout << "Contravariant␣3D␣basis␣+␣O3␣=␣"
                          << G3 << endl;
   //
   MATRIX contra_g2(2, num_eyes, num_points);
   //
   contra_g2 = PERSP.transformation(G3, a_distance_G, num_points);
   //
   if (ERROR) std::cout <<
       "Image␣coordinates␣of␣the␣contravariant␣basis␣vectors:␣"
```

```
                              << contra_g2 << endl;
        // ==============================================================
    //
    // Determine minimum and maximum window coordinates of the image
    // ==============================================================
    //
        double min1, min2, max1, max2;
        // PERSP.fenster(min1, min2, max1, max2);
           PERSP.window (min1, min2, max1, max2);
        //
        if (ERROR) std::cout << "Minimum␣window␣coordinates:␣min1=␣"
                      << min1 << "␣min2=␣" << min2 << endl;
        if (ERROR) std::cout << "Maximum␣window␣coordinates:␣max1=␣"
                      << max1 << "␣max2=␣" << max2 << std::endl;
        //
        std::cout << endl;
        std::cout << "End␣of␣IMAGE␣Coordinates" << endl;
        std::cout << "=========================" << endl << endl;
        //
        // ****************************
        ERROR = false;
        // ****************************
// ==============================================================
// Display basis vectors on the 2D-projection screen graphically
// ==============================================================
        int draw_basis = 0;
        long color = 0;
        double diff, diff1, diff2, fact, fact1, fact2;
        strstream buffer1;
        //
        // Image size
        double image1, image2;
        image1 = 1920.0 / 53.3;
        image2 = 1080.0 / 36.0;
        // ###################
        diff1 = max1 - min1;
        diff2 = max2 - min2;
        // Reduction factors to actual size in [cm]
        fact1 = diff1 / image1;
        fact2 = diff2 / image2;
        // ---------------------------------------------------------
        // For parallel perspective evtl. reduce to actual size
        if (type_persp == 0)
        {
         if ((diff1 > image1) || (diff2 > image2))
         {
           std::cout << "The␣image␣area␣is:␣" << image1 << "␣*␣"
                     << image2 << '\n';
           std::cout << "The␣image␣size␣is:␣" << diff1 << "␣*␣"
                     << diff2 << '\n';
           fact = image1 / diff1;
           diff = image2 / diff2;
           if (diff < fact) fact = diff;
           //
```

```
        std::cout <<
         "Shrink␣the␣image?␣Input:␣Scaling␣factor␣(suggestion=␣";
        //
        // preciseness
        std::cout.width(5);
        //*** std::cout.precision(2);
        //*** std::cout.setf(ios::fixed);
        //
        std::cout << fact << "):" << std::endl;
        std::cout <<
            "Change?␣->␣1␣=␣yes␣/␣0␣=␣no␣(default):␣" << std::endl;
        std::cout <<
            "========================================";
        std::cin >> input;              outFILE << input << std::endl;
        if (input == 1)
        {
           std::cout << "Enter␣fact:" << std::endl;
           std::cin >> fact;               outFILE << fact << std::endl;
           input = 0;
        }
        fact1 = fact1 * fact;
        fact2 = fact2 * fact;
      };
     };
      //
      // Destructor -> delete constructor
      // =================================
      delete &PERSP;
// ################# End of PERSPEKTIVE ###########################
//                    ==================
// ================================================================
// Constructor call for Python graphics
// ==================================
// Image scaling and evtl. drawing of circles arond points
//                                               -> dimension [mm]
   double circle = 1.0;
// -.-.-.-.-.-.-.-.-.-.
// Constructor
   PyPLOT& PYTHON = *new PyPLOT(fact1, fact2, circle);
// ================================================================

// Foreground colors                               X11/CSS4 - Colors
// =================                               ========   exact
// 0=_BLACK   = k   8=_GRAY
// 1=_BLUE    = b   9=_LIGHTBLUE
// 2=_GREEN   = g  10=_LIGHTGREEN    g not exactly green   -> #008000
// 3=_CYAN    = c  11=_LIGHTCYAN     c not exactly cyan    -> #00FFFF
// 4=_RED     = r  12=_LIGHTRED
// 5=_MAGENTA = m  13=_LIGHTMAGENTA  m not exactly magenta -> #FF00FF
// 6=_YELLOW  = y  14=_LIGHTYELLOW   y not exactly yellow  -> #FFFF00
// 7=_WHITE   = w  15=_BRIGHTWHITE
// ================================================================
 // Set the "super title"
 // if (type_persp == 0) PYTHON.bild_titel ("Parallel Perspective");
```

```
  if (type_persp == 0) PYTHON.image_title("Parallel␣Perspective");
 else
 {
// if(num_eyes == 1) PYTHON.bild_titel ("Central Perspective");
   if(num_eyes == 1) PYTHON.image_title("Central␣Perspective");
// if(num_eyes == 2) PYTHON.bild_titel ("Stereo Central Perspective");
   if(num_eyes == 2) PYTHON.image_title("Stereo␣Central␣Perspective");
 };
 //
 // Show plot
 // *** PYTHON.zeige_bild();
 // *** PYTHON.show_image();
// ================================================================
// Horizon - draw eye point
// ========================
// Set foreground color
// color = 6;                                        // -> yellow
//
   std::string M_p;
   std::string F_, L_l, L_s, L_p;
//
   M_p = "o";               // Node
   F_ = "y";  L_p = ":";    // Line
//
   if (num_eyes == 1) { F_ = "red";  L_p = ":"; };
// ------------------------------------------------
//
// H - A draw and label
// ====================
// PYTHON.linie(a2, h2, 0, F_, L_p, "H");
   PYTHON.line (a2, h2, 0, F_, L_p, "H");
//
// Show plot
//*** PYTHON.zeige_bild();
//*** PYTHON.show_image();
//
// A -> Label with graphic text
// (Circle around A and distance to the eye point)
// ===============================================
//
// PYTHON.punkt(a2, 0, F_, M_p, "A", a0_spacing_A);
   PYTHON.point(a2, 0, F_, M_p, "A", a0_spacing_A);
//
// Show plot
//*** PYTHON.zeige_bild();
//*** PYTHON.show_image();
//
// Left eye foreground color -> cyan
// =========================
// color = 3;                                           // -> cyan
// F_ = "c";                                // c is not exactly CYAN !!!
//
   M_p = "o";                                           // Node
   F_ = "cyan";  L_l = "-";  L_s = "--";  L_p = ":";  // Line
```

```
// ------------------------------------------------------------------
//
// Loop over the number of eyes
// ============================
for (eyes = 0; eyes < num_eyes; eyes++)
{
    // ==============================================================
    // Display objects with graphic text
    // Set foreground color
    // color = 0;                                          // -> black
       if (num_eyes == 1) F_ = "k";
    //
    // Origin U of the position vector O
    // =================================
    // *** PYTHON.punkt(u2, eyes, F_, M_p, " ");
    // *** PYTHON.point(u2, eyes, F_, M_p, " ");
    //
    // Show plot
    //*** PYTHON.zeige_bild();
    //*** PYTHON.show_image();
    //
    // Tip 0 of the position vector O
    // ==============================
    //*** PYTHON.punkt(o2, eyes, F_, M_p, "0");
    //*** PYTHON.point(o2, eyes, F_, M_p, "0");
    //
    // Show plot
    //*** PYTHON.zeige_bild();
    //*** PYTHON.show_image();
    //
    //
    // ==============================================================
    // Draw basis vectors
    // ==================
    // Set forground color
    // color = 0;                                          // -> black
       if (num_eyes == 1)
       {
           F_ = "k";  L_l = "-";  L_s = "--";  L_p = ":";  // line
       };
    //
    // ==============================================================
    // Draw Cartesian coordinate system
    // ================================
    // PYTHON.dreibein(u2, x2, eyes, F_, L_l, "E");
       PYTHON.tripod  (u2, x2, eyes, F_, L_l, "E");
    //
    // Show plot
    //*** PYTHON.zeige_bild();
    //*** PYTHON.show_image();
    //
    //
    // Draw position vector (Line coordinate origin - 0)
    // =================================================
```

```
    // PYTHON.linie(o2, u2, eyes, F_, L_p, " ");
       PYTHON.line (o2, u2, eyes, F_, L_p, "␣");
    //
    // Show plot
    //*** PYTHON.zeige_bild();
    //*** PYTHON.show_image();
    //
    // Draw covariant or/and contravariant basis
    // ==========================================
    if (eyes == 0)
    {
       std::cout << "Draw␣basis␣vectors␣in␣one␣image?" << endl;
       std::cout <<"1␣=␣covariant␣/␣0␣=␣both␣/␣-1␣=␣contravariant:␣"
                 << endl;
       std::cout <<"===============================================␣";
       std::cin >> draw_basis;         outFILE << draw_basis << endl;
       std::cout << endl;
    };
    //
    // Draw covariant basis vectors
    // ============================
    //
    if (draw_basis >= 0)
    // PYTHON.dreibein(o2, cov_g2, eyes, F_, L_s, "G");
       PYTHON.tripod  (o2, cov_g2, eyes, F_, L_s, "G");
    //
    // Show plot
    //*** PYTHON.zeige_bild();
    //*** PYTHON.show_image();
    //
    // Draw contravariant basis vectors
    // ================================
    //
    if (draw_basis <= 0)
    // PYTHON.dreibein(o2, contra_g2, eyes, F_, L_l, "g");
       PYTHON.tripod  (o2, contra_g2, eyes, F_, L_l, "g");
    //
    // Show plot
    //*** PYTHON.zeige_bild();
    //*** PYTHON.show_image();
    //
    // ==============================================================
    // Right eye foreground color
    //
    // color = 4;                                           // -> red
       M_p = "o";                                         // Node
       F_ = "red";  L_l = "-";  L_s = "--";  L_p = ":";  // Line
    // ------------------------------------------------------------
}  // End of loop over eyes
// ================================================================
// Show plot
// PYTHON.zeige_bild();
   PYTHON.show_image();
// Pause **********************************************************
```

```
std::cout << endl;
std::cout <<
        "New␣eye␣position?␣->␣1␣=␣yes␣/␣0␣=␣no␣(default):␣" << endl;
std::cout <<
        "================================================␣";
   std::cin >> input;                              outFILE << input << endl;
// ******************************************************************
// Screen clear
// ============
// Destructor -> delete constructor
// ===============================
delete &PYTHON;
// ############ End: PyPlot #########################################
//
    if (input != 1)
    {
        std::cout << "Graphics␣output␣skipped␣!" << endl;
        type_persp = -1;
    };
} // End: type_persp ======= End Block: Type of Perspective =========
// Jump back to new perspective
// ===========================
   std::cout <<
  "Generate␣new␣perspective␣?␣->␣1␣=␣yes␣/␣0␣=␣no␣(default):" <<endl;
   std::cout <<
  "=========================================================␣";
   std::cin >> input;                              outFILE << input << endl;
   if (input == 1)  z_persp = 1; else z_persp = 0;
   //
} while (z_persp == 1);  // End: z_persp
// =============== End of Block: new Perspective ====================
// End of Graphic Mode -> close the Python instance
//
// *** std::cout << "FINALIZE" << endl;
   Py_Finalize();
};
// =================================================================
    std::cout << endl << "End␣BASES_3D" << endl;
    std::cout <<         "============" << endl << endl;
    // ==============================================================
    std::cout << "PAUSE:␣" << endl;
    std::cout << "======␣";
    std::cin >> input;                             outFILE << input << endl;
    // ==============================================================
return 2;
};
// =================================================================
```

As a demonstration example for the application of the function BASES_3D, the example of the function METRIC (Listing 4.3) from Section 4.6.2 is referred to, which spans a tetrahedron in three-dimensional Euclidean space. This tetrahedron is later used in Section 7.2 for the example of the triangle element (see Fig. 7.2).

The following is the input and output protocol of the BASES_3D function. It illustrates the dialog for determining the viewing direction and the horizon. The subsequent process involves the coordinate transformation using the methods of the class PERSPEKTIVE. Then, the 2D representation of the basis vectors on the projection screen is created using the methods of the class PyPLOT.

Listing 5.4: Numerical Example of the Graphics Classes PERSPEKTIVE and PyPLOT

```
3D-Perspective of the Basis Vectors
===================================

Skip 3D-perspective -> exit program?
-> 1 = yes / 0 = no (default):
============================== 0

Enter type of perspective
1 = Central perspective / 0 = Parallel perspective / -1 = end:
============================================================= 1
Perspective: The eye A is directed towards the vanishing point F,
the horizon point H determines the horizontal line and,
spans by SPACING the projection screen, perpendicular to A - F !

Enter eye point A:
Default coordinates for A:
[ (1*3) * (1*1) ]-MATRIX:
Columns 0 to 2
 8.000e+00 -4.500e+01  1.500e+01

Pause -> enter a character:

Change? -> 1 = yes / 0 = no (default):
====================================== 0

Enter vanishing point F (in center of object):
Default coordinates F:
[ (1*3) * (1*1) ]-MATRIX:
Columns 0 to 2
 5.000e+00  5.000e+00  5.000e+00

Pause -> enter a character:

Change? -> 1 = yes / 0 = no (default):
====================================== 0

Enter horizon point H (determines the horizon):
Default coordinates H:
[ (1*3) * (1*1) ]-MATRIX:
Columns 0 to 2
 1.500e+01  1.600e+01  5.000e+00

Pause -> enter a character:

Change? -> 1 = yes / 0 = no (default):
```

```
====================================== 0
For a better overview,
             the basis vectors are shifted by the position vector
------------------------------------------------------------------

Enter endpoint of position vector O
Default coordinates O:
[ (1*3) * (1*1) ]-MATRIX:
Columns 0 to 2
 2.500e+00  2.500e+00  2.500e+00

Pause -> enter a character:

Change? -> 1 = yes / 0 = no (default):
====================================== 0

Enter number of eyes: 1 = Mono / 2 = Stereo:
============================================ 2

SPACING from the eye point A to the reference point 0
                             of the projection screen:
61.2588


Optical stretching of the contravariant and Cartesian bases
by the FACTOR = 3.68403

Change? -> 1 = yes / 0 = no (default):
================================================ 0

Pause -> enter a character:
End of IMAGE Coordinates
=========================

Draw basis vectors in one image?
1 = covariant / 0 = both / -1 = contravariant:
============================================== 0

New eye position? -> 1 = yes / 0 = no (default):
================================================ 0

Generate new perspective ? -> 1 = yes / 0 = no (default):
======================================================== 0

End BASES_3D
============
```

The subsequent Figure 5.5 displays the result of the function BASES_3D as a central perspective on the covariant and contravariant basis vectors. When viewed with red/cyan glasses, the colored basis vectors emerge vividly from the projection screen. This spatial view in Euclidean space is especially informative for understanding the position and arrangement of the basis vectors and their angular orientation to each other. Similarly, other vectors, lines, and wireframe bodies may also be integrated into the BASES_3D function.

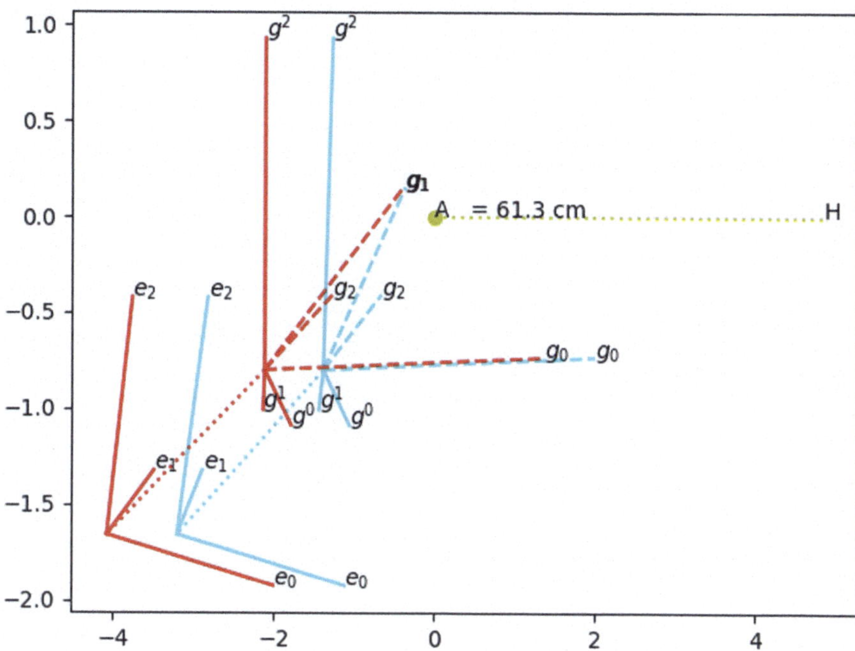

Fig. 5.5: Central perspective of covariant and contravariant basis vectors with optical stretching of Cartesian and contravariant bases by a factor of 3.68

Internet Links

Meissner 2023. Meißner, U. F.: Object-oriented Tensor and Matrix Classes (2023)
`https://rebrand.ly/Matrix_Classes`

Microsoft 2023. Visual Studio - Visualizing data with mathplotlib (2023)
`https://learn.microsoft.com/en-us/search/?terms=matplotlib-cpp-visual-studio`
`https://stackoverflow.com/questions/66507618/running-python-matplotlibcpp-in-visual-studio-2019`

Chapter 6
Tensor Analysis for Finite Elements

Abstract

In this chapter, the fundamentals of tensor analysis are compiled for use within the finite element method. Similar to the classical approaches such as [Klingbeil 1966], the differential geometry of the spatial bodies is treated first, followed by the mechanical fundamentals for the deformation of elastic continua. The tensor representations are then specialized to the finite element approximations for the modeling of the geometry and the description of the displacement fields. These explanations follow the classical compendia, such as [Zienkiewicz/Taylor 1989] or even the textbook [Meissner/Maurial 2000]. Here, however, they are consistently presented in the preferred index notation for matrices and tensors. Derived exemplary for numerical purposes are the matrix relations for the approximate spanning of the geometry of individual elements and for their stiffness relations under external loading as well as for the calculation of internal forces. Since the focus is primarily on the principle of matrix methodology, the plane continua of the plate theory are selected as a relatively transparent field of application. This is sufficiently complex, but nevertheless illustrative, in order to demonstrate the advantages of the new matrix methods with typical examples.

In the previous chapter, the coupling of the coordinates of the Euclidean point space with a position vector bound to the Cartesian reference system was established with the introduction of the Euclidean vector space. So far, linear and orthogonal coordinates x^i have been considered. In the following, this perspective is, on the one hand, generalized to the treatment of curvilinear, so-called natural coordinates, and on the other hand, to the consideration of vector fields, which can span a variety of geometric bodies depending on such coordinates.

The primary goal is to prepare the known methods of differential geometry and surface theory accordingly and, using tensor analysis, make them accessible to the finite element method to the extent that numerical methods can be derived and specified. Regarding further considerations, refer to the extensive literature [Zienkiewicz/Taylor 1989], [Clough/Penzien 1975] and [Bathe 1986].

U. F. Meissner, *Tensor Calculus with Object-Oriented Matrices for Numerical Methods in Mechanics and Engineering*, https://doi.org/10.1007/978-3-031-59302-4_6

6.1 Differential Geometry with Curvilinear Coordinates

In the Euclidean vector space $\mathbb{R}^3$, according to (4.1.1) with reference to the spatial Cartesian basis vectors $\mathbf{e}_i$, the position vector $\mathbf{x}$

$$\overline{PO} := \mathbf{x} = x^i * \mathbf{e}_i$$

was introduced, identifying the position of each point $P(x^i)$ in space. Specific geometric bodies representing lines, surfaces, or volumes in space are now to be formed from a subset of these points in $\mathbb{R}^3$.

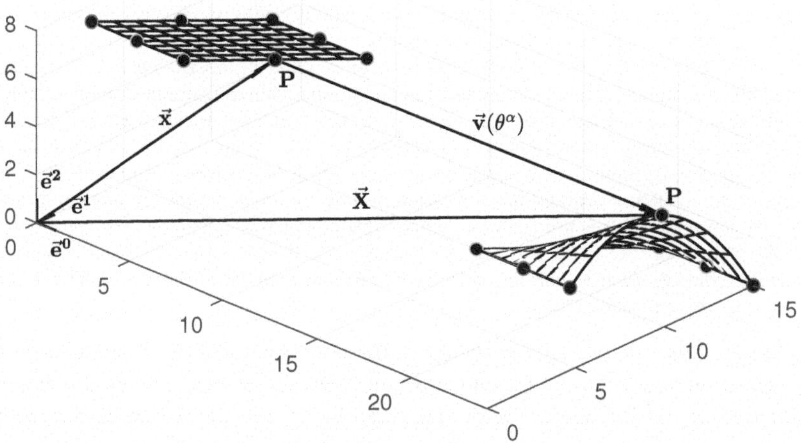

Fig. 6.1: Deformation of an undeformed body by the displacement field $\mathbf{v}(\theta^\alpha)$ depending on the body-fixed coordinates θ^α

To identify the corresponding body points, so-called ***natural coordinates***, also referred to as ***body-fixed coordinates*** $\boldsymbol{\theta^\alpha}$ ***or*** $\boldsymbol{\lambda^\alpha}$, are introduced, which are typically dimensionless and restricted to normalized intervals, such as $\theta^\alpha \in \mathbb{R}\ [-1,+1]$ or $\lambda^\alpha \in \mathbb{R}\ [0,+1]$. These coordinates are understood to be firmly connected to the body and deform convectively, so that even after deformation, the body points remain uniquely identifiable. In continuum mechanics, this connection for each point P is described by extending the position vector $\mathbf{x}$ of the undeformed body by the displacement vector $\mathbf{v}$, so that the position vector $\mathbf{X}$

$$\boxed{\mathrm{X}^i = x^i + v^i(\theta^\alpha) \quad \Leftrightarrow \quad \mathbf{X} = \mathbf{x} + \mathbf{v}(\theta^\alpha)} \tag{6.1.1}$$

identifies the points of the deformed body (cf. Fig. 6.1).

The relationship between the spatial and the body-fixed coordinates can, in each specific case, be accomplished by a special coordinate transformation of the type

$$\boxed{\mathrm{x}^i = f^i(\theta^\alpha)}\,, \tag{6.1.2}$$

called ***shape function*** in the finite element method. This transformation determines the shape of the respective body based on the body-specific parameters θ^α through the vector field f^i, which locates the body's own points with x^i and thus spans the body in space.

Following surface theory, it is simplifying assumed that only smooth bodies are involved, which do not exhibit invaginations, folds, penetrations, or similar degenerations, so that the mapping is bijective. Accordingly, the functions f^i must be continuous and, with respect to curvature, differentiable twice. For the boundaries of the bodies, only regular surfaces and lines are considered, which can be captured by (6.1.2).

To illustrate the presented considerations, the common example of the ***globe*** according to Fig. 6.2 is chosen as a model. Spherical coordinates are used as natural coordinates. These are the spherical coordinates θ^0 and θ^1, as well as the coordinate θ^2 in the direction of the radius as

$$\begin{array}{ll} \textit{Longitude} & \varphi = \theta^0 \in \mathbb{R}\,[-\pi, +\pi]\,, \\ \textit{Latitude} & \phi = \theta^1 \in \mathbb{R}\,[> -\pi/2, < +\pi/2]\,, \\ \textit{Radius} & r = \theta^2 \in \mathbb{R}\,[> 0, R]\,. \end{array}$$

The ***coordinate transformation of spherical coordinates*** is carried out by mapping body-fixed θ^α onto spatial coordinates x^i

$$\mathrm{x}^i = \begin{pmatrix} r * cos(\phi) * cos(\varphi) \\ r * cos(\phi) * sin(\varphi) \\ r * sin(\phi) \end{pmatrix} = \begin{pmatrix} \theta^2 * cos(\theta^1) * cos(\theta^0) \\ \theta^2 * cos(\theta^1) * sin(\theta^0) \\ \theta^2 * sin(\theta^1) \end{pmatrix}. \tag{6.1.3}$$

With a constant radius $\theta^2 = R$, the position vector x^i describes the surface of the sphere. The Gaussian surface parameters θ^0 and θ^1 then represent the ***polar coordinates*** of the sphere's surface and form a two-dimensional Euclidean subspace $\mathbb{R}^2$ on this surface. If, furthermore, θ^0 is held constant, the position vector with variable θ^1 on the sphere's surface represents the ***meridians*** - the great circles of longitude - and in the reverse case with constant θ^1 and variable θ^0, the ***parallels*** of latitude arranged orthogonally to them as displayed.

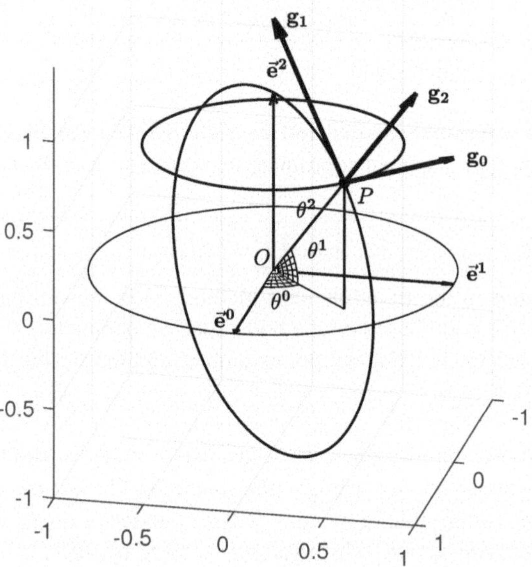

Fig. 6.2: Spherical coordinates θ^α of the globe

With the tangents to such pairs of convective coordinate lines, corresponding tangent planes can be spanned at each point of the sphere, formed by the associated ***covariant basis vectors*** $\mathbf{g}_\alpha$:

$$\boxed{\begin{array}{ll} g^i{}_\alpha = x^i{}_{,\alpha} & \Leftrightarrow \quad \mathbf{g}_\alpha = \mathbf{x}_{,\alpha} \\ \textit{with the partial derivatives} & \\ x^i{}_{,\alpha} = \partial x^i / \partial \theta^\alpha & \Leftrightarrow \quad \mathbf{x}_{,\alpha} = \partial \mathbf{x} / \partial \theta^\alpha \\ \textit{in common abbreviation of the tensor notation} & \end{array}} \,. \qquad (6.1.4)$$

For the example, the following ***covariant basis vectors of the sphere*** result from the corresponding differentiation of (6.1.3) , within the spherical body:

$$g^i{}_\alpha = \begin{pmatrix} -r*\cos(\theta^1)*\sin(\theta^0) & -r*\sin(\theta^1)*\cos(\theta^0) & \cos(\theta^1)*\cos(\theta^0) \\ r*\cos(\theta^1)*\cos(\theta^0) & -r*\sin(\theta^1)*\sin(\theta^0) & \cos(\theta^1)*\sin(\theta^0) \\ 0 & r*\cos(\theta^1) & \sin(\theta^1) \end{pmatrix} . \qquad (6.1.5)$$

For the ***covariant metric tensor of the sphere***, as discussed in Section 4.2, the following is obtained at each point using (4.5.1):

$$g_{\alpha\beta} = g^i{}_\alpha * g^i{}_\beta = \begin{pmatrix} r^2 * \cos^2(\theta^0) & 0 & 0 \\ 0 & r^2 & 0 \\ 0 & 0 & 1 \end{pmatrix} . \tag{6.1.6}$$

Therefore, the covariant basis vectors are perpendicular to each other at the intersection points of the spherical coordinates, as expected on the surface of the globe.

With the determinant $g = det(g_{\alpha\beta})$, the ***volume element of the sphere*** is given by

$$\sqrt{g} = r^2 * \cos(\theta^0) \ . \tag{6.1.7}$$

By inverting (6.1.6), according to (4.2.4), the ***contravariant metric tensor of the sphere*** is obtained:

$$g^{\alpha\beta} = 1/r^2 \begin{pmatrix} 1/\cos^2(\theta^0) & 0 & 0 \\ 0 & 1 & 0 \\ 0 & 0 & r^2 \end{pmatrix} , \tag{6.1.8}$$

from which, according to (4.2.6), the ***contravariant basis vectors of the sphere***

$$g_i{}^\alpha = 1/r * \begin{pmatrix} -\sin(\theta^0)/\cos(\theta^1) & -\sin(\theta^1) * \cos(\theta^0) & r * \cos(\theta^1) * \cos(\theta^0) \\ \cos(\theta^0)/\cos(\theta^1) & -\sin(\theta^1) * \sin(\theta^0) & r * \cos(\theta^1) * \sin(\theta^0) \\ 0 & \cos(\theta^1) & r * \sin(\theta^1) \end{pmatrix} \tag{6.1.9}$$

are derived.

After obtaining the tangent vector space of the basis $\mathbf{g}_\alpha$ with (6.1.5) ff, the curvature measure of the sphere is briefly addressed. For this purpose, as mentioned above, the surface associated with a constant radius $r = R$ is considered.

According to (6.1.5), the ***covariant normal vector*** on this surface has components

$$g^i{}_2 = \left(\cos(\theta^1) * \cos(\theta^0) \ \cos(\theta^1) * \sin(\theta^0) \ \sin(\theta^1) \right) ,$$

and the two-dimensional ***contravariant basis of the sphere surface*** is given, according to (6.1.9), by

$$\mathrm{G}_i{}^\alpha = 1/R * \begin{pmatrix} -\cos(\theta^1) * \sin(\theta^0) & -\sin(\theta^1) * \cos(\theta^0) \\ \cos(\theta^1) * \cos(\theta^0) & -\sin(\theta^1) * \sin(\theta^0) \\ 0 & \cos(\theta^1) \end{pmatrix} .$$

Thus, according to (6.1.7), the
1st fundamental form of surface theory

$$\sqrt{G} = R^2 * \cos(\theta^0)$$

is represented by the ***area element of the sphere surface***.

The basic idea for measuring the curvature of the surface is to determine how quickly, measured as an angular change, the gradient of the normal vector, perpendicular to the surface, changes in its position relatively to the tangential plane of the two-dimensional basis vectors. This can be determined by the scalar product of these two vector quantities as the
2nd fundamental form of surface theory

$$\boxed{\begin{array}{l} b^{\alpha}{}_{\beta} = -G_i{}^{\alpha} * g^i{}_{2},_{\beta} \\ \textit{or due to the orthogonality} \,(4.2.7)\;\; (G_i{}^{\alpha} * g^i{}_2),_{\beta} = 0 \\ b^{\alpha}{}_{\beta} = \;G_i{}^{\alpha},_{\beta} * g^i{}_2 \end{array}} \quad \textit{with}\; \alpha,\beta \in \mathbb{N}\,[0,1] \qquad (6.1.10)$$

and the ***curvature tensor*** $b^{\alpha}{}_{\beta}$.

For the example of the sphere, the gradient of the covariant normal vector is:

$$g^i{}_{2},_{\beta} = \begin{pmatrix} -\cos(\theta^1) * \sin(\theta^0) & -\sin(\theta^1) * \cos(\theta^0) \\ \cos(\theta^1) * \cos(\theta^0) & -\sin(\theta^1) * \sin(\theta^0) \\ 0 & \cos(\theta^1) \end{pmatrix}.$$

Thus, the ***curvature tensor of the sphere surface***

$$b^{\alpha}{}_{\beta} = -\begin{pmatrix} 1/R & 0 \\ 0 & 1/R \end{pmatrix} \qquad (6.1.11)$$

is gained, as expected, with the same curvature measure in both orthogonal directions.

Of general interest are surfaces whose curvatures are different in diverse directions. This property is further investigated by determining the eigenvalues of the curvature tensor $b^{\alpha}{}_{\beta}$. The determinant of the eigenvalue problem

$$\begin{vmatrix} b^0{}_0 - \lambda & b^0{}_1 \\ b^1{}_0 & b^1{}_1 - \lambda \end{vmatrix} = 0$$

with the two ***invariants of the curvature tensor***

$$\boxed{\begin{array}{ll} \textit{mean curvature} & 2H = b^{\alpha}{}_{\alpha} \\ \textit{Gaussian curvature} & K = \det(b^{\alpha}{}_{\beta}) \end{array}} \qquad (6.1.12)$$

leads to the characteristic equation $\lambda^2 - 2H\lambda + K = 0$
with the eigenvalues $\lambda_{\alpha} = H \pm \sqrt{H^2 - K}$
and the ***principal curvatures***

$$\boxed{\begin{aligned} 1/R_\alpha &= \lambda_\alpha \\ 2H &= 1/R_0 + 1/R_1 \\ K &= 1/R_0 * 1/R_1 \end{aligned}}\,. \tag{6.1.13}$$

The ***Gaussian curvature measure*** thus describes

$$\boxed{\begin{array}{l} \textit{with} \\ K > 0 \textit{ the } \textbf{\textit{ellipsoid}} \textit{ with elliptical curvatures,} \\ K = 0 \textit{ the } \textbf{\textit{paraboloid}} \textit{ with parabolic curvatures,} \\ K < 0 \textit{ the } \textbf{\textit{hyperboloid}} \textit{ with hyperbolic curvatures.} \end{array}} \tag{6.1.14}$$

6.2 Covariant Derivative and Symbolic Operators

In (6.1.2), the functional relationship between the spatial Cartesian coordinates x^i and the body-fixed coordinates θ^α was already presented. Now, the scalar field

$$I(\theta^\alpha) = I(x^i)$$

is considered as a tensor of 0th order, and its partial derivative

$$\partial I/\partial\theta^\alpha = \partial I/\partial x^i \ \partial x^i/\partial\theta^\alpha$$

is examined due to tensor properties. Using (6.1.4), it follows that ***the first derivative of a scalar is a tensor of the 1st order***

$$\boxed{\begin{array}{l} I,_\alpha = I,_i * g^i{}_\alpha \\ \text{or} \\ I,_i = I,_\alpha * g_i{}^\alpha \end{array}}\ , \tag{6.2.1}$$

because it transforms according to the transformation property (4.3.2).

In (6.1.10), the partial derivative of the contravariant basis vectors was formed, and now, the partial derivatives of the covariant and contravariant basis vectors are examined due to tensor properties.
This is done through the orthogonality (4.2.8):

$$\delta^\gamma{}_\alpha = g_i{}^\gamma * g^i{}_\alpha$$

$$\delta_\gamma{}^\alpha = g^i{}_\gamma * g_i{}^\alpha .$$

The partial derivative of the δ-matrix

$$\delta_\gamma{}^\alpha,_\beta = (g^i{}_\gamma * g_i{}^\alpha),_\beta \equiv 0$$

$$g^i{}_\gamma,_\beta * g_i{}^\alpha = -g^i{}_\gamma * g_i{}^\alpha,_\beta$$

also explains the change of sign in the Christoffel symbols listed following in (6.2.2).

With the above relations, the two ***partial derivatives of the basis vectors*** can now be related to the respective tensor bases $g_i{}^\gamma$ and $g^i{}_\gamma$ using the ***Christoffel symbols of the 2nd kind***:

$$\boxed{\begin{array}{l} g_i{}^\alpha,_\beta = -\Gamma^\alpha{}_{\beta\gamma} * g_i{}^\gamma \\ \textit{with the Christoffel symbol } \Gamma^\alpha{}_{\beta\gamma} = -g_i{}^\alpha,_\beta * g^i{}_\gamma \\ \text{and} \\ g^i{}_\alpha,_\beta = \ \Gamma^\gamma{}_{\alpha\beta} * g^i{}_\gamma \\ \textit{with the Christoffel symbol } \Gamma^\gamma{}_{\alpha\beta} = \ g^i{}_\alpha,_\beta * g_i{}^\gamma \end{array}}\ . \tag{6.2.2}$$

These Christoffel symbols are symmetric regarding the indices α *and* β due to the derivatives of the basis vectors, and they vanish in rectilinear coordinates. It is crucial to note that they behave like 1st-order tensors only concerning the index γ, as clarified by particular transformations with the basis vectors $g_i{}^\gamma$ *and* $g^i{}_\gamma$.

Similarly, the ***Christoffel symbol of the 1st kind*** can be obtained by multiplying it with the metric tensor:

$$\boxed{\Gamma_{\gamma\alpha\beta} = g_{\gamma\lambda} * \Gamma^{\lambda}{}_{\alpha\beta}} \,. \tag{6.2.3}$$

According to [Klingbeil 1966], it is possible to express the ***Christoffel symbol of the 2nd kind*** solely through the covariant metric tensor and its partial derivatives:

$$\boxed{\Gamma^{\gamma}{}_{\alpha\beta} = 1/2 * g^{\gamma\lambda} * (g_{\lambda\alpha},_{\beta} + g_{\lambda\beta},_{\alpha} - g_{\alpha\beta},_{\lambda}) \quad \text{(without proof)}} \,. \tag{6.2.4}$$

Through the Christoffel symbols, the explanation leads logically to the covariant derivatives for all tensors of higher order. As displayed by the following differentiation of the transformation (4.3.2)

$$v^i,_k = g_k{}^\beta * v^i,_\beta = g_k{}^\beta * (g^i{}_\alpha * v^\alpha),_\beta = g_k{}^\beta * (g^i{}_\alpha * v^\alpha,_\beta + g^i{}_{\alpha},_{\beta} * v^\alpha),$$

it becomes obvious that even the partial derivative of a 1st-order tensor does not have tensor properties. However, by subsequently using the corresponding Christoffel symbol (6.2.2)

$$v^i,_k = g^i{}_\alpha * g_k{}^\beta * (v^\alpha,_\beta + \Gamma^\alpha{}_{\beta\gamma} * v^\gamma) = g^i{}_\alpha * g_k{}^\beta * v^\alpha|_\beta$$

the ***transformation properties of a 2nd-order tensor*** can be restored by introducing the mathematical construct of the covariant derivative

$$\boxed{\begin{aligned} v^i,_k &= g^i{}_\alpha * g_k{}^\beta * v^\alpha|_\beta \\ v_i,_k &= g_i{}^\alpha * g_k{}^\beta * v_\alpha|_\beta \end{aligned}} \,, \tag{6.2.5}$$

where the same procedure applies to the partial derivative of the covariant vector $v_i,_k$.

From this modification, the ***covariant derivatives of 1st-order tensors*** result as follows:

$$\boxed{\begin{aligned} v^\alpha|_\beta &= v^\alpha,_\beta + \Gamma^\alpha{}_{\beta\gamma} * v^\gamma \\ v_\alpha|_\beta &= v_\alpha,_\beta - \Gamma^\gamma{}_{\alpha\beta} * v_\gamma \end{aligned}} \,. \tag{6.2.6}$$

Respectively, tensors of higher order transform for each derivative index according (4.4.2), if the partial derivative is consequently substituted by the covariant derivative, as defined. While the complete differential of a vector field v^α was previously formed in rectilinear coordinates with the partial derivative

$$dv^\alpha = v^\alpha,_\beta \, d\theta^\beta,$$

it now transforms into the ***total differential*** in curvilinear coordinates using the covariant derivative:

$$\boxed{\mathrm{d}v^{\alpha} = v^{\alpha}|_{\beta}\, d\theta^{\beta}} \,. \tag{6.2.7}$$

By this formulation, the tensor properties of the derivative keep ensured.

Finally, the ***symbolic operators***, being also related to the covariant derivative, are briefly presented in comparison with the preferred index notation.

First, the ***Nabla operator*** ∇ is listed:

$$\boxed{\begin{array}{ll} \textit{Gradient of a scalar field} & \\ \varphi = \nabla w = \textbf{grad}\, w & \Leftrightarrow \quad \varphi_{\alpha} = w_{,\alpha} \\ \textit{Gradient of a vector field} & \\ \varepsilon = \nabla v = \textbf{grad}\, v & \Leftrightarrow \quad \epsilon_{\alpha\beta} = v_{\alpha}|_{\beta} \end{array}} \,, \tag{6.2.8}$$

$$\boxed{\begin{array}{ll} \textit{Divergence of a vector field} & \\ q = \nabla v = \textbf{div}\, v & \Leftrightarrow \quad q = v^{\alpha}|_{\alpha} \\ \textit{Divergence of a tensor field} & \\ p = \nabla \cdot N = \textbf{div}\, N & \Leftrightarrow \quad p^{\alpha} = N^{\alpha\beta}|_{\beta} \end{array}} \,, \tag{6.2.9}$$

$$\boxed{\begin{array}{ll} \textit{Rotation of a vector field} & \\ r = \nabla \times v = \textbf{rot}\, v & \Leftrightarrow \quad r^{\alpha} = \varepsilon^{\alpha\beta\gamma} * v_{\gamma}|_{\beta} \end{array}} \,. \tag{6.2.10}$$

Next, the ***Laplace-Operator*** Δ is presented:

$$\boxed{\begin{array}{ll} \textit{Potential equation of a scalar field} & \\ q = \Delta h = \nabla \cdot \nabla h & \Leftrightarrow \quad q = h|_{\alpha}{}^{\alpha} = h|^{\alpha}{}_{\alpha} \\ \textit{Potential equation of a vector field} & \\ s = \Delta v = \nabla \cdot \nabla v & \Leftrightarrow \quad s^{\alpha} = v^{\alpha}|_{\beta}{}^{\beta} \\ \textit{Bipotential equation of a scalar field} & \\ q = \Delta\Delta w = \nabla \cdot \nabla \nabla \cdot \nabla w & \Leftrightarrow \quad q = w|_{\alpha}{}^{\alpha}{}_{\beta}{}^{\beta} \end{array}} \,. \tag{6.2.11}$$

Arithmetic combinations of symbolic operators and their specific rules, which can be found in the literature, are not discussed here, as they are generally unnecessary when using index notation.

6.3 Integral Theorems of Spatial Continua

First, we consider the ***Gauss/Green's integral theorem*** for a spatial tensor field $\boldsymbol{T}$

$$\boxed{\iiint_{\Omega} T^{\alpha}|_{\beta}\, d\Omega = \iint_{\Gamma} T^{\alpha} * n_{\beta}\, d\Gamma}\,, \tag{6.3.1}$$

where, for instance, a volume integral can be converted into a closed surface integral, with n_{β} representing the outward directed normal vector on the surface.

In continuum mechanics, this integral theorem is of central importance when applying the principle of virtual displacements or the principle of minimum potential energy. The ***virtual strain energy*** of elastic continua is then represented as the product of virtual strains $\delta v_{\alpha}|_{\beta}$ and internal stresses $\sigma^{\alpha\beta}$.
Using Gauss/Green's theorem, it is possible to eliminate the covariant derivative of virtual displacements δv_{α} on the left side of the following equation and transform the volume integral into a different and more advanced form:

$$\boxed{\delta A^{(i)} = \iiint_{\Omega} \delta v_{\alpha}|_{\beta} * \sigma^{\alpha\beta}\, d\Omega = \iint_{\Gamma} \delta v_{\alpha} * \sigma^{\alpha\beta} * n_{\beta}\, d\Gamma - \iiint_{\Omega} \delta v_{\alpha} * \sigma^{\alpha\beta}|_{\beta}\, d\Omega}\,. \tag{6.3.2}$$

This solution approach is based on a boundary value problem in which the surface of the continuum Γ is divided into the regions Γ_{v}, where displacements $\overline{v}_{\alpha}$ are prescribed, and Γ_{σ}, where external surface forces $\overline{p}^{\alpha}$ are active, because different ***boundary conditions*** must be implemented there:

$$\boxed{\begin{array}{ll} \delta v_{\alpha} = 0 \text{ due to } v_{\alpha} = \overline{v}_{\alpha} & \text{at } \Gamma_{v} \\ \sigma^{\alpha\beta} * n_{\beta} = \overline{p}^{\alpha} & \text{at } \Gamma_{\sigma} \\ \text{with} & \Gamma = \Gamma_{v} \cap \Gamma_{\sigma} \end{array}}\,. \tag{6.3.3}$$

Together with the additionally external body forces $\overline{q}^{\alpha}$ acting on the continuum, the ***virtual work of external forces*** sums up to:

$$\boxed{\delta A^{(a)} = -\iint_{\Gamma_{\sigma}} \delta v_{\alpha} * \overline{p}^{\alpha}\, d\Gamma - \iiint_{\Omega} \delta v_{\alpha} * \overline{q}^{\alpha}\, d\Omega}\,. \tag{6.3.4}$$

The ***principle of virtual displacements*** requires that the total virtual work of the elastic continuum vanishes:

$$\boxed{\begin{array}{l} \delta A = \delta A^{(i)} + \delta A^{(a)} = 0, \\ \textit{and it leads to the condition} \\ -\iiint_{\Omega} \delta v_{\alpha} * (\sigma^{\alpha\beta}|_{\beta} + \overline{q}^{\alpha})\, d\Omega + \iint_{\Gamma_{\sigma}} \delta v_{\alpha} * (\sigma^{\alpha\beta} * n_{\beta} - \overline{p}^{\alpha})\, d\Gamma = 0 \end{array}}\,. \tag{6.3.5}$$

Due to the arbitrariness of virtual displacements δv_{α}, this results in the ***equilibrium conditions of the continuum***, described by the differential equation

$$\boxed{\sigma^{\alpha\beta}|_\beta + \overline{q}^\alpha = 0} \,, \tag{6.3.6}$$

and in the ***stress boundary conditions*** according to (6.3.3) at the region Γ_σ where surface forces act. The integral over the region Γ_v, where displacements are prescribed along the supports, vanishes because the virtual displacement $\delta\overline{v}_\alpha = 0$ is zero by definition.

In conclusion of these considerations, it is necessary to refer briefly to the ***integration of tensorial functions over multidimensional domains*** with curvilinear coordinates, which is important in mechanics for evaluating integral work principles and finite element residual matching methods.
Generally, domain integrals of the form

$$\boxed{\begin{array}{ll} \iiint_V f(\theta^0,\theta^1,\theta^2)\, dV, & where\ dV = \sqrt{g} * d\theta^0 * d\theta^1 * d\theta^2 \\ \iint_A f(\theta^0,\theta^1)\, dA, & where\ dA = \sqrt{g} * d\theta^0 * d\theta^1 \\ \int_L f(\theta^0)\, dL, & where\ dL = \sqrt{g} * d\theta^0 \end{array}} \tag{6.3.7}$$

are to be treated, where the determination of the volume, area, and line elements $\sqrt{g}$ has already been discussed according to (4.2.3) and depends on the particular dimension.

6.4 Discretization with Finite Elements

Section 6.1 has already discussed the relationship between spatial and body-fixed coordinates, demonstrating the generation of geometric bodies using shape functions (6.1.2) with the example of the Globe.

In the following, this approach will be exemplarily deepened for the Finite Element Method using some typical examples. The goal is primarily to present the typical tensor methods necessary for modeling mechanical bodies and their deformations, and to implement them directly in object-oriented numerical methods using index-based matrix calculus.
However, the procedure will be limited to some typical geometric shapes to highlight their characteristics. For the transfer and extension of this approach to the comprehensive variety of common Finite Element Methods, standard works in the field, such as [Zienkiewicz/Taylor 1989], [Clough/Penzien 1975], [Bathe 1986], or [Baker 1991], are available for various applications.

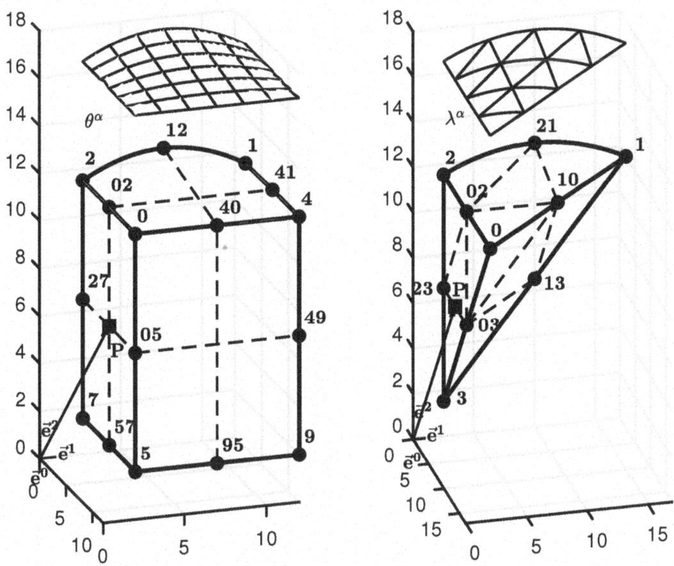

Fig. 6.3: Spatial hexahedral and tetrahedral element with natural coordinate lines in the interval $\theta^\alpha \in \mathbb{R}[-1,+1]$ and $\lambda^\alpha \in \mathbb{R}[0,+1]$, respectively.

To illustrate these connections, two typical basic elements, a ***hexahedron*** and a ***tetrahedron***, are shown in Fig. 6.3. These are volume elements, whose body points $P(\theta^\alpha)$ in three dimensions with $\alpha \in \mathbb{N}[0,2]$ and $P(\lambda^\alpha)$ with $\alpha \in \mathbb{N}[0,3]$ are measured using body-fixed, quasi 'carved-in' natural coordinates. These points deform with the deformation of the elements. In the two- and one-dimensional cases, such elements degenerate into corresponding surface and line elements.

Since the finite element method is an approximation method, geometry in Euclidean vector space $\mathbb{R}^3$ is generally spanned according to (6.1.2) by the ***position vector*** X^i ***of the finite element*** in terms of the position vectors $X^i_{\textcircled{K}}$ of the nodal points $\textcircled{K}$ and by the polynomial shapes of the shape functions $N_{\textcircled{K}}(\theta^\alpha)$.

$$\boxed{X^i = X^i_{\textcircled{K}} * N_{\textcircled{K}}(\theta^\alpha)} \, . \tag{6.4.1}$$

The finite element discretization of a complex overall structure is usually carried out by a large number of different but structurally identical finite line, surface, or volume elements, which are geometrically and mechanically compatible at their respective boundaries, connected without gaps.

6.4.1 Geometry of the Quadrilateral Element

To span and describe the geometry, two different quadrilateral elements are chosen here as typical patterns of the hexahedron.

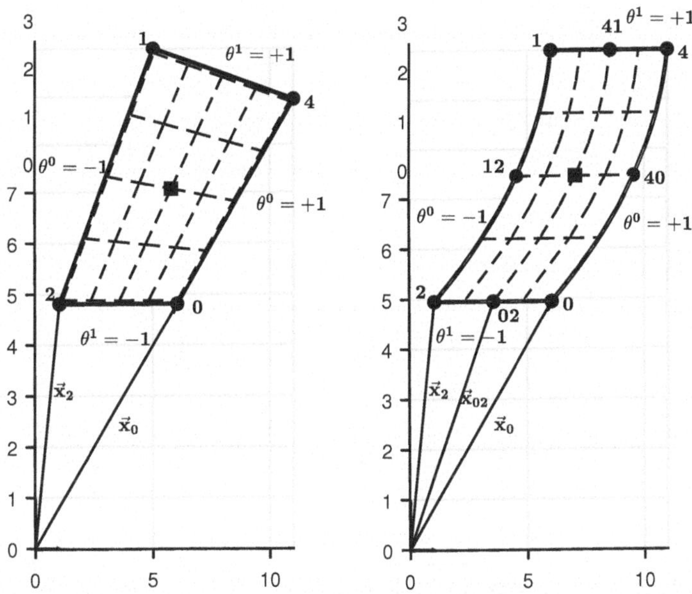

Fig. 6.4: Bilinear Lagrange quadrilateral and quadratic Serendipity quadrilateral with natural coordinates $\theta^\alpha \in \mathbb{R}[-1,+1]$

Fig. 6.4 illustrates the example of straight and curved quadrilaterals, where the body points are measured using the natural coordinates θ^α with $\alpha \in \mathbb{N}[0,1]$. For the sake of clarity, it is assumed in the following that the $\mathbf{e}_0 - \mathbf{e}_1$-plane of the Cartesian coordinates x_i with $i \in \mathbb{N}[0,1]$ is parallel to the plane of the quadrilaterals.

bilinear Quadrilateral

The ***four shape functions of the 'Lagrange' family*** forming the bilinear quadrilateral element are given by:

$$\begin{aligned} N_{0} &= 1/4*(1+\theta^0)*(1-\theta^1) \\ N_{1} &= 1/4*(1-\theta^0)*(1+\theta^1) \\ N_{2} &= 1/4*(1-\theta^0)*(1-\theta^1) \\ N_{4} &= 1/4*(1+\theta^0)*(1+\theta^1) \end{aligned} \tag{6.4.2}$$

with corner nodes in the order $\textcircled{K} \in \mathbb{N}\ [0,1,2,4]$.

quadratic Quadrilateral

For the quadratic quadrilateral element of the 'Serendipity' family without mid-nodes, the corresponding ***eight shape functions*** are:

$$\boxed{\begin{array}{ll} N_{\textcircled{0}} = 1/4*(1+\theta^0)*(1-\theta^1)*(\ \theta^0-\theta^1-1) & N_{\textcircled{02}} = 1/2*(1-\theta^0*\theta^0)*(1-\theta^1) \\ N_{\textcircled{1}} = 1/4*(1-\theta^0)*(1+\theta^1)*(-\theta^0+\theta^1-1) & N_{\textcircled{40}} = 1/2*(1-\theta^1*\theta^1)*(1+\theta^0) \\ N_{\textcircled{2}} = 1/4*(1-\theta^0)*(1-\theta^1)*(-\theta^0-\theta^1-1) & N_{\textcircled{41}} = 1/2*(1-\theta^0*\theta^0)*(1+\theta^1) \\ N_{\textcircled{4}} = 1/4*(1+\theta^0)*(1+\theta^1)*(\ \theta^0+\theta^1-1) & N_{\textcircled{12}} = 1/2*(1-\theta^1*\theta^1)*(1-\theta^0) \end{array}} \tag{6.4.3}$$

with nodes in the order $\textcircled{K} \in \mathbb{N}\ \{0,1,2,4,02,40,41,12\}$.

These shape functions have the property that they each have the value $N_{\textcircled{K}}(\theta^0_{\textcircled{K}}, \theta^1_{\textcircled{K}}) = 1$ at node $\textcircled{K}$ and the value $N_{\textcircled{K}}(\theta^0_{\textcircled{I}}, \theta^1_{\textcircled{I}}) = 0$ at all other nodes $\textcircled{I}$.

Using these shape functions, according to (6.1.4), the ***covariant basis vectors*** can be determined at each point $P(\theta^\alpha)$ as tangents to the natural coordinate lines

$$\boxed{g^i{}_\alpha = X^i{}_{,\alpha} = X^i_{\textcircled{K}} * N_{\textcircled{K}},_\alpha} \ . \tag{6.4.4}$$

For the ***quadratic quadrilateral***, the ***partial derivatives of the shape functions*** are given by:

$$N_{\textcircled{K}},_\alpha = 1/4* \begin{pmatrix} +(1-\theta^1)*(+2*\theta^0-\theta^1) & -(1+\theta^0)*(-2*\theta^1+\theta^0) \\ -(1+\theta^1)*(-2*\theta^0+\theta^1) & +(1-\theta^0)*(+2*\theta^1-\theta^0) \\ -(1-\theta^1)*(-2*\theta^0-\theta^1) & -(1-\theta^0)*(-2*\theta^1-\theta^0) \\ +(1+\theta^1)*(+2*\theta^0+\theta^1) & +(1+\theta^0)*(+2*\theta^1+\theta^0) \\ -4*\theta^0*(1-\theta^1) & -2*(1-\theta^0*\theta^0) \\ +2*(1-\theta^1*\theta^1) & -4*\theta^1*(1+\theta^0) \\ -4*\theta^0*(1+\theta^1) & +2*(1-\theta^0*\theta^0) \\ -2*(1-\theta^1*\theta^1) & -4*\theta^1*(1-\theta^0) \end{pmatrix} , \tag{6.4.5}$$

and the ***partial derivatives of the shape functions*** for the ***bilinear quadrilateral*** are given by:

$$N_{\textcircled{K}},_\alpha = 1/4* \begin{pmatrix} +(1-\theta^1) & -(1+\theta^0) \\ -(1+\theta^1) & +(1-\theta^0) \\ -(1-\theta^1) & -(1-\theta^0) \\ +(1+\theta^1) & +(1+\theta^0) \end{pmatrix} . \tag{6.4.6}$$

The position of the covariant basis vectors is shown in Fig. 6.5.

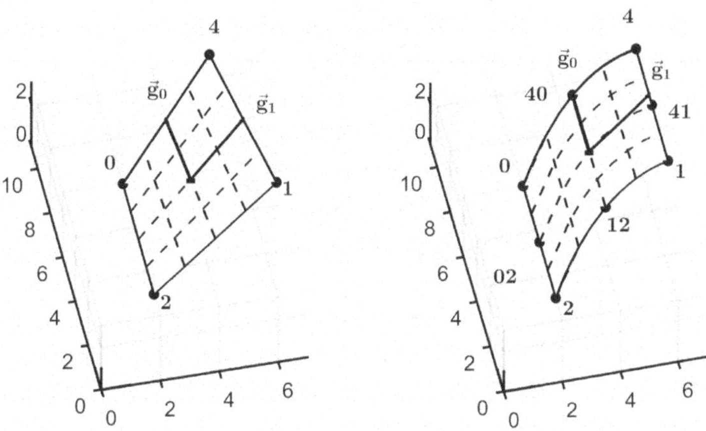

Fig. 6.5: Covariant basis vectors $\mathbf{g}_\alpha$ of the quadrilateral elements depending on the natural coordinates θ^α

As noticeable from the complexity of the polynomial terms, it is generally not possible to determine all geometric base quantities analytically. This is especially true for the formation of the inverse matrix for the contravariant measure tensor $g^{\alpha\beta}$ and also for the square root of the area element $\sqrt{g}$. For specific support points of $\theta^\alpha_{\textcircled{G}}$, numerical matrix methods must be used to locally determine and process the values of such quantities. This approach will be described in more detail later with examples. At this point, however, the chosen examples are simplified enough to allow the typical procedure to be traced analytically.

Parallelogram

The bilinear quadrilateral is therefore simplified to a parallelogram element, whose opposite sides are aligned parallel, and that with its ***covariant basis vectors***

$$\boxed{g^i{}_\alpha = 1/2 * \left(X^i_{\textcircled{0}} - X^i_{\textcircled{2}} \;\; X^i_{\textcircled{1}} - X^i_{\textcircled{2}} \right)} \tag{6.4.7}$$

is shown in Fig. 6.6.

The ***covariant metric*** (4.2.2) results in the lengths l_α and the angle α between the basis vectors:

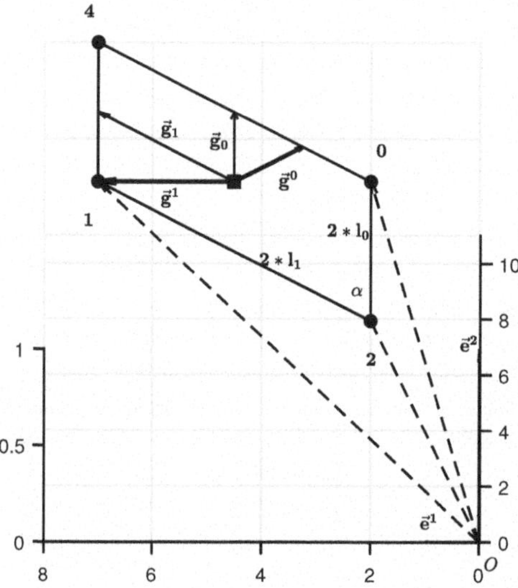

Fig. 6.6: Covariant basis vectors $\mathbf{g}_\alpha$ and contravariant basis vectors $\mathbf{g}^\alpha$ (enlarged by factor A/4) of the parallelogram element

$$g_{\alpha\beta} = g^i{}_\alpha * g^i{}_\beta = \begin{pmatrix} (l_0)^2 & l_0 * l_1 * cos(\alpha) \\ l_0 * l_1 * cos(\alpha) & (l_1)^2 \end{pmatrix} \tag{6.4.8}$$

with $l_0 = \sqrt{g^i{}_0 * g^i{}_0} = \sqrt{g_{00}}$ and $l_1 = \sqrt{g^i{}_1 * g^i{}_1} = \sqrt{g_{11}}$,
and the ***area element*** $\sqrt{g} = l_0 * l_1 * sin(\alpha) = A/4$, one quarter of the element area.

The ***contravariant metric*** is usually calculated as the inverse matrix of the covariant metric $g_{\alpha\beta}$. However, here it can also be obtained directly for analytical determination using the tensorbasis transformation (4.5.14) by rotating the axes by 90° and swapping the components:

$$\begin{aligned} g^{\alpha\beta} &= \varepsilon^{\alpha\gamma} * g_{\gamma\delta} * \varepsilon^{\beta\delta} \\ &= 1/g \begin{pmatrix} (l_1)^2 & -l_0 * l_1 * cos(\alpha) \\ -l_0 * l_1 * cos(\alpha) & (l_0)^2 \end{pmatrix} \end{aligned} . \tag{6.4.9}$$

In this easy way, the ***contravariant basis*** can also be determined directly by rotating the covariant basis (6.4.7) using the tensorbasis transformation (4.5.14) again:

$$\begin{aligned} g_i{}^\alpha &= e_{ik} * g^k{}_\gamma * \varepsilon^{\alpha\gamma} \\ &= 1/\sqrt{g} * \begin{pmatrix} l_1 * sin(\alpha) & 0 \\ -l_1 * cos(\alpha) & l_0 \end{pmatrix} = \begin{pmatrix} 1/l_0 & 0 \\ -cot(\alpha)/l_0 & 1/(l_1 * sin(\alpha)) \end{pmatrix} \end{aligned} . \tag{6.4.10}$$

The basis vectors are depicted in Fig. 6.6. It is important to note that the contravariant basis vectors are perpendicular to the covariant basis vectors, and the length of the basis vector $\mathbf{g}^1$ depends solely on the height $h = 1/l_1 * \sin(\alpha)$ of node ① above the baseline ⓪-②. This relationship will also apply later to (6.4.23) for the linear triangular element.

6.4.2 Geometry of the Triangle Element

As additional examples, two triangular elements are chosen here as typical basic forms of the tetrahedron, as displayed in Fig. 6.7:
- a linearly bounded triangle with 3 nodes Ⓚ $\in \mathbb{N}\,[0,2]$
and
- a quadratically curved triangle with 6 nodes Ⓚ $\in \mathbb{N}\,\{0,1,2,10,21,02\}$.
It is assumed below that the $\mathbf{e}_0 - \mathbf{e}_1$ plane of the Cartesian coordinates x^i with $i \in \mathbb{N}[0,1]$ is parallel to the triangle plane.

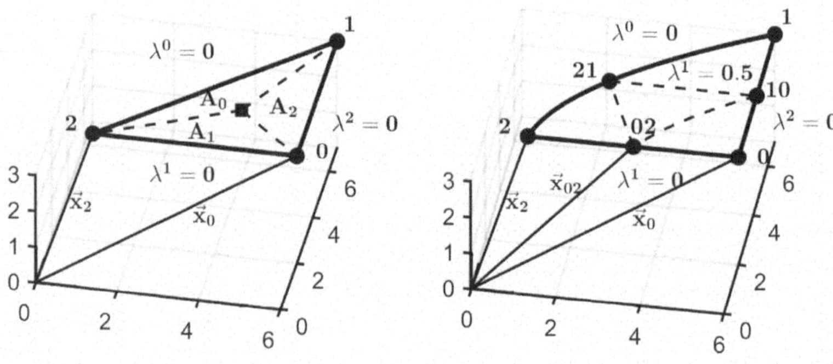

Fig. 6.7: Linear and quadratic triangular element with natural coordinates $\lambda^\alpha \in \mathbb{R}[0,+1]$

To identify the body points, body-fixed triangle area coordinates, or briefly refered to as ***area coordinates***, are used in this case for symmetry reasons as natural coordinates in the normalized interval $\lambda^\alpha \in \mathbb{R}\,[0,+1]$. These are defined by partial areas A_α (see Fig. 6.7), which are assigned to the corner nodes,

$$\boxed{\begin{array}{l}\lambda^\alpha = A_\alpha / A \\ \textit{and which are linearly dependent on each other due to } \sum_\alpha A_\alpha = A \\ \lambda^0 + \lambda^1 + \lambda^2 = 1\end{array}} \quad . \qquad (6.4.11)$$

This two-dimensional case of area coordinates λ^α with $\alpha \in \mathbb{N}\,[0,2]$ can also be generalized to ***volume coordinates of spatial tetrahedral elements*** $\lambda^0 + \lambda^1 + \lambda^2 + \lambda^3 = 1$ with four sub-volumes V^α and $\alpha \in \mathbb{N}\,[0,3]$ according to [Zienkiewicz/Taylor 1989].

It is helpful to interpret the area model discussed here as a special case of such a three-dimensional tetrahedron, with its $\mathbf{e}_0 - \mathbf{e}_1$-base at a height of $x^2 = 1$ above the apex of the coordinate origin $\circledcirc$. This virtual model may support the clarity of the following considerations:

linear Triangle

According to (6.4.1), the ***three shape functions of the 'Lagrange' family*** spanning the linear triangle element are:

$$\boxed{\begin{aligned} N_{(0)} &= \lambda^0 \\ N_{(1)} &= \lambda^1 \\ N_{(2)} &= \lambda^2 \end{aligned}}\,. \tag{6.4.12}$$

quadratic Triangle

For the quadratic triangle element of the ***'Lagrange' family, the six shape functions*** are as follows:

$$\boxed{\begin{aligned} N_{(0)} &= \lambda^0 * (2*\lambda^0 - 1) & N_{(10)} &= 4*\lambda^0*\lambda^1 \\ N_{(1)} &= \lambda^1 * (2*\lambda^1 - 1) & N_{(21)} &= 4*\lambda^1*\lambda^2 \\ N_{(2)} &= \lambda^2 * (2*\lambda^2 - 1) & N_{(02)} &= 4*\lambda^2*\lambda^0 \end{aligned}}\,. \tag{6.4.13}$$

These shape functions also have the property that they take at node (K) the value $N_{(K)}(\lambda^0_{(K)}, \lambda^1_{(K)}, \lambda^2_{(K)}) = 1$ and at all other nodes (I) the value $N_{(K)}(\lambda^0_{(I)}, \lambda^1_{(I)}, \lambda^2_{(I)}) = 0$.

With these approaches, according to (6.1.4), the ***covariant basis vectors*** for the natural coordinate lines can be determined at every point $P(\lambda^\alpha)$:

quadratic Triangle

For the quadratic triangle, the ***partial derivative of the shape functions*** is given by

$$N_{(K)},_\alpha = \begin{pmatrix} 4*\lambda^0 - 1 & 0 & 0 \\ 0 & 4*\lambda^1 - 1 & 0 \\ 0 & 0 & 4*\lambda^2 - 1 \\ 4*\lambda^1 & 4*\lambda^0 & 0 \\ 0 & 4*\lambda^2 & 4*\lambda^1 \\ 4*\lambda^2 & 0 & 4*\lambda^0 \end{pmatrix}. \tag{6.4.14}$$

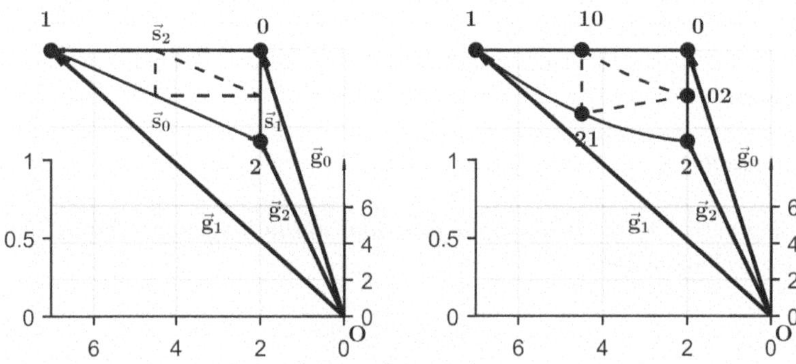

Fig. 6.8: Covariant basis vectors $\mathbf{g}_\alpha$ of triangle elements in dependence of natural coordinates λ^α

linear Triangle

The ***partial derivative of the shape functions*** for the linear triangle is

$$\mathrm{N}_{\textcircled{K}},_\alpha = \begin{pmatrix} 1 & 0 & 0 \\ 0 & 1 & 0 \\ 0 & 0 & 1 \end{pmatrix} = \delta_{\textcircled{K}\alpha} \,. \tag{6.4.15}$$

This results in the ***covariant basis vectors of the linear triangular element*** at the apex of the virtual tetrahedron:

$$\boxed{\mathrm{g}^i{}_\alpha = x^i_{\textcircled{\alpha}} = \left(x^i_{\textcircled{0}}\; x^i_{\textcircled{1}}\; x^i_{\textcircled{2}}\right) = \begin{pmatrix} x^0_{\textcircled{0}} & x^0_{\textcircled{1}} & x^0_{\textcircled{2}} \\ x^1_{\textcircled{0}} & x^1_{\textcircled{1}} & x^1_{\textcircled{2}} \end{pmatrix}} \,. \tag{6.4.16}$$

Alternatively, in the nodes $\textcircled{K}$ of the triangle, if needed, another ***set of covariant basis vectors*** $\mathbf{s}_\alpha$ ***along the adjacent triangle sides*** may be defined, whose position and orientation are also shown in Fig. 6.8:

$$\begin{array}{ll} \overset{\textcircled{0}}{\mathbf{s}}_1 = \mathbf{g}_1 - \mathbf{g}_0 & \overset{\textcircled{0}}{s}{}^i{}_1 = g^i{}_1 - g^i{}_0 = (x^i{}_{\textcircled{1}} - x^i{}_{\textcircled{0}}) \\ \overset{\textcircled{1}}{\mathbf{s}}_2 = \mathbf{g}_2 - \mathbf{g}_1 & \overset{\textcircled{1}}{s}{}^i{}_2 = g^i{}_2 - g^i{}_1 = (x^i{}_{\textcircled{2}} - x^i{}_{\textcircled{1}}) \\ \overset{\textcircled{2}}{\mathbf{s}}_0 = \mathbf{g}_0 - \mathbf{g}_2 & \overset{\textcircled{2}}{s}{}^i{}_0 = g^i{}_0 - g^i{}_2 = (x^i{}_{\textcircled{0}} - x^i{}_{\textcircled{2}}) \end{array} \tag{6.4.17}$$

with linear dependence

$$\overset{\textcircled{0}}{\mathbf{s}}_1 + \overset{\textcircled{1}}{\mathbf{s}}_2 + \overset{\textcircled{2}}{\mathbf{s}}_2 = 0 \qquad \overset{\textcircled{0}}{s}{}^i{}_1 + \overset{\textcircled{1}}{s}{}^i{}_2 + \overset{\textcircled{2}}{s}{}^i{}_0 = 0\,.$$

To determine the contravariant basis vectors, the covariant basis (6.4.16) in the sense of the virtual tetrahedron is supplemented with a third row, to capture the linear dependence of the triangle coordinates $\lambda^0 + \lambda^0 + \lambda^0 = 1$ during the coordinate transformation:

$$a^I = a^I{}_\alpha * \lambda^\alpha$$

$$\text{with } a^I = \begin{pmatrix} x^0 \\ x^1 \\ 1 \end{pmatrix} \text{ and } a^I{}_\alpha = \begin{pmatrix} x^0{}_{\textcircled{0}} & x^0{}_{\textcircled{1}} & x^0{}_{\textcircled{2}} \\ x^1{}_{\textcircled{0}} & x^1{}_{\textcircled{1}} & x^1{}_{\textcircled{2}} \\ 1 & 1 & 1 \end{pmatrix} \text{ and } \lambda^\alpha = \begin{pmatrix} \lambda^0 \\ \lambda^1 \\ \lambda^2 \end{pmatrix}. \tag{6.4.18}$$

The ***area element of the triangle*** is then obtained according to (2.2.9) as:

$$\boxed{\begin{array}{l} \sqrt{g} = det(a^I{}_\alpha) = e_{IKL} * a^I{}_0 * a^K{}_1 * a^L{}_2 \\ \quad = 2 * (A_0 + A_1 + A_2) = 2 * A \\ \textit{twice the element area,} \\ \textit{and the base areas } A_\alpha \textit{ of the virtual pyramid} \\ \quad 2 * A_0 = x^0{}_{\textcircled{1}} * x^1{}_{\textcircled{2}} - x^0{}_{\textcircled{2}} * x^1{}_{\textcircled{1}} \\ \quad 2 * A_1 = x^0{}_{\textcircled{2}} * x^1{}_{\textcircled{0}} - x^0{}_{\textcircled{0}} * x^1{}_{\textcircled{2}} \\ \quad 2 * A_2 = x^0{}_{\textcircled{0}} * x^1{}_{\textcircled{1}} - x^0{}_{\textcircled{1}} * x^1{}_{\textcircled{0}} \end{array}}\,. \tag{6.4.19}$$

This is followed by the covariant metric of the virtual tetrahedron with

$$\boxed{\mathrm{a}_{\alpha\beta} = a^I{}_\alpha * a^I{}_\beta}\,. \tag{6.4.20}$$

Finally, the inversion of the coordinate transformation (6.4.18)

$$\lambda^\beta = a_I{}^\beta * a^I$$

yields the mixed-variant identity matrix

$$\boxed{\mathrm{a}_\alpha{}^\beta = a^I{}_\alpha * a_I{}^\beta = \delta_\alpha{}^\beta}\,, \tag{6.4.21}$$

from which the contravariant basis vectors of the virtual cylinder result as the inverse matrix of $a^I{}_\alpha$

$$a_I{}^\beta = \frac{1}{2 * A} \begin{pmatrix} \beta_0 & \beta_1 & \beta_2 \\ \alpha_0 & \alpha_1 & \alpha_2 \\ 2 * A_0 & 2 * A_1 & 2 * A_2 \end{pmatrix}.$$

Alternatively, the contravariant basis vectors can also be derived directly from the covariant basis vectors $a^L{}_\delta$ using the tensorbasis transformation (4.5.15):

$$\boxed{\mathrm{a}_I{}^\beta = 1/2 * e_{ILN} * a^L{}_\delta * a^N{}_\lambda * \varepsilon^{\delta\lambda\beta}} \,. \tag{6.4.22}$$

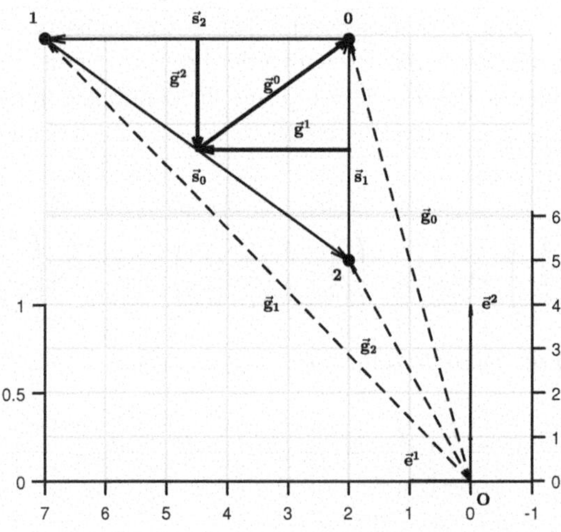

Fig. 6.9: Contravariant basis vectors $\mathbf{g}^\alpha$ of the linear triangular element (enlarged by a factor of A)

The ***contravariant basis vectors of the linear triangular element*** are obtained by finally removing the third row of the matrix $a_I{}^\beta$:

$$\boxed{\begin{array}{l} g_i{}^\beta = \frac{1}{2*A}\begin{pmatrix} \beta_0 & \beta_1 & \beta_2 \\ \alpha_0 & \alpha_1 & \alpha_2 \end{pmatrix} \\[1em] \textit{with} \quad \begin{array}{l} \alpha_0 = x^0_{\textcircled{2}} - x^0_{\textcircled{1}} \\ \alpha_1 = x^0_{\textcircled{0}} - x^0_{\textcircled{2}} \\ \alpha_2 = x^0_{\textcircled{1}} - x^0_{\textcircled{0}} \end{array} \qquad \begin{array}{l} \beta_0 = -(x^1_{\textcircled{2}} - x^1_{\textcircled{1}}) \\ \beta_1 = -(x^1_{\textcircled{0}} - x^1_{\textcircled{2}}) \\ \beta_2 = -(x^1_{\textcircled{1}} - x^1_{\textcircled{0}}) \end{array} \\[1em] \textit{with the vector lengths} \quad \begin{array}{ll} g^{00} = (s_0)^2/g & s_0 = |x^i_{\textcircled{2}} - x^i_{\textcircled{1}}|/2A \\ g^{11} = (s_1)^2/g & s_1 = |x^i_{\textcircled{0}} - x^i_{\textcircled{2}}|/2A \\ g^{22} = (s_2)^2/g & s_2 = |x^i_{\textcircled{1}} - x^i_{\textcircled{0}}|/2A. \end{array} \\[1em] \textit{and the linear dependence} \\ g_i{}^0 + g_i{}^1 + g_i{}^2 = 0 \;\; \text{resp.} \;\; \mathbf{g}^0 + \mathbf{g}^1 + \mathbf{g}^2 = 0 \end{array}} \,. \tag{6.4.23}$$

Their arrangement is shown in Fig. 6.9.

As already noticed in the case of the parallelogram (6.4.10), determining the contravariant basis vectors $\mathbf{g}^0, \mathbf{g}^1, \mathbf{g}^2$ is independent of the specific vertex.
From the labels, it is evident that, however, they are assigned to the respective sides of the triangle. On these sides, they are perpendicular and point inward toward the opposite nodes. To avoid confusion in the labels, it is therefore useful to designate the triangle sides according to the opposite corner nodes $\textcircled{K}$ with s_K (cf. Fig. 6.8).

In a purely two-dimensional view of the planar triangle element, three equivalent but different reference systems for capturing the associated tensorial quantities arise at the corner nodes, as illustrated in Fig. 6.9 with $\mathbf{s}_\alpha$:

- ***at corner 0*** with coordinates λ^1 *and* λ^2, the covariant basis vectors ${}^{\textcircled{0}}\mathbf{g}_1 = \mathbf{s}_2$ and ${}^{\textcircled{0}}\mathbf{g}_2 = -\mathbf{s}_1$, as well as the contravariant basis vectors $\mathbf{g}^1$ *and* $\mathbf{g}^2$,

- ***at corner 1*** with coordinates λ^2 *and* λ^0, the covariant basis vectors ${}^{\textcircled{1}}\mathbf{g}_2 = \mathbf{s}_0$ and ${}^{\textcircled{1}}\mathbf{g}_0 = -\mathbf{s}_2$, as well as the contravariant basis vectors $\mathbf{g}^2$ *and* $\mathbf{g}^0$,

- ***at corner 2*** with coordinates λ^0 *and* λ^1, the covariant basis vectors ${}^{\textcircled{2}}\mathbf{g}_0 = \mathbf{s}_1$ and ${}^{\textcircled{2}}\mathbf{g}_1 = -\mathbf{s}_0$, as well as the contravariant basis vectors $\mathbf{g}^0$ *and* $\mathbf{g}^1$.

These designations and those of the associated tensorial quantities can also be easily generated by cyclically interchanging the index values $\alpha = 0, 1, 2$ one after the other.

Due to the transformation (6.4.23)

$$t^0 + t^1 + t^2 = t^i * (g_i{}^0 + g_i{}^1 + g_i{}^2) = 0$$

the ***contravariant tensor components***

$$\begin{array}{ll} t^0 + t^1 + t^2 = 0 & t_0{}^0 + t_0{}^1 + t_0{}^2 = 0 \\ t^{00} + t^{10} + t^{20} = 0 & t^0{}_0 + t^1{}_0 + t^2{}_0 = 0 \end{array} \qquad (6.4.24)$$

are also ***linearly dependent*** on each other.

For the transformation of covariant and contravariant tensor components, the ***contravariant metric of the linear triangular element*** is obtained according to (6.4.22):

$$\boxed{g^{\alpha\beta} = a_I{}^\alpha * a_I{}^\beta} \, . \qquad (6.4.25)$$

In connection with the described triangular elements, it is important to note that the evaluation of the ***domain integrals*** (6.3.7) can be greatly simplified ***due to the natural*** volume, surface, and line ***coordinates***. According to [Zienkiewicz/Taylor 1989], the practical integration rule ***for straight-edged triangles***

$$\boxed{\begin{array}{ll} \iiint_V f[(\lambda^0)^i,(\lambda^1)^k,(\lambda^2)^l,(\lambda^3)^m]\,dV = \frac{i!\,k!\,l!\,m!}{(i+k+l+m+3)!} * \sqrt{g}, & \sqrt{g} = 2V \\ \iint_A f[(\lambda^0)^i,(\lambda^1)^k,(\lambda^2)^l)]\,dA = \frac{i!\,k!\,l!}{(i+k+l+2)!} * \sqrt{g}, & \sqrt{g} = 2A \\ \int_L f[(\lambda^0)^i,(\lambda^1)^k)]\,dL = \frac{i!\,k!}{(i+k+1)!} * \sqrt{g}, & \sqrt{g} = 2L \end{array}} \tag{6.4.26}$$

is applicable for the exact integration of the polynomial terms with $\sqrt{g}$ according to (4.2.3).

6.5 Least Squares Approximation with Finite Elements

The ultimate goal of tensor analysis for finite elements is to discuss the approaches of the numerical approximation method in this section. This method forms the basis for the application of object-oriented matrix methods to two typical finite elements. The geometry of the quadrilateral and triangular elements has already been discussed for both examples. Now, the variational principle (6.3.5) of virtual strain energy explained in Section 6.1 is referred to.

Since the 1960s, the method of finite elements has created a highly flexible numerical approximation for various physical applications, using polynomial approximations based on Taylor series. This method fundamentally minimizes generally unknown truncation errors between the exact solution of the systems of differential equations and their approximate solutions [Meissner/Maurial 2000].

Two fundamental questions arise initially: a) what measure is used to assess the residue of truncation errors, and b) what is the mathematical-mechanical method that minimizes the residual. In the case of linear systems of differential equations and when potential problems are present, the solution approach can be clearly described, taking the example of the elastic continuum. The present example is further simplified by reducing the three-dimensional continuum to a ***two-dimensional plate structure***. To achieve this, a narrow two-dimensional slice of thickness t is cut out, which can only carry loads in the direction of its plane.

With the residual $\tilde{E}$ between the exact and approximate solutions of the elastic plate problem, the following ***least squares functional*** is formulated as an error norm according to [Meissner 1996]:

$$\boxed{\begin{aligned}\tilde{E} &= 1/2 * \iint_A (v_\alpha|_\beta - \hat{v}_\alpha|_\beta) * E^{\alpha\beta\gamma\delta} * (v_\delta|_\gamma - \hat{v}_\delta|_\gamma)\, dA \\ &= 1/2 * \iint_A \tilde{v}_\alpha|_\beta * E^{\alpha\beta\gamma\delta} * \tilde{v}_\delta|_\gamma\, dA \;\geq 0\end{aligned}} \quad with\, \alpha,\beta \in \mathbb{N}\,[0,1]. \tag{6.5.1}$$

Here, v_α represents the unknown exact displacement field of the mechanical problem, and $\hat{v}_\alpha$ its FEM approximate solution. To estimate the error order O of the ***truncation error*** $\tilde{v}$ ***of the Taylor series*** of the approximation polynomials $\hat{v}_\alpha$

$$\boxed{\tilde{v}_\alpha = v_\alpha - \hat{v}_\alpha = O(h^{p+1})}\,, \tag{6.5.2}$$

the definition of a characteristic measure h, as (4.2.3) for geometric discretization, is generally required, cf. (6.4.8) and (6.4.19), as well as p for the highest order of the used complete power series. With the mth derivative of the displacement $\hat{v}_\alpha$ the power of the error order generally decreases by the degree m to $O(h^{p+1-m})$. For $v_\alpha|_\beta$ the order $\tilde{v}_\alpha|_\beta = O(h^{p+1-1})$ holds accordingly.

The minimum of $\tilde{E}$ will be numerically determined in the following. It is assumed that the constitutive ***law of elastic material***

$$\boxed{\mathrm{n}^{\alpha\beta} = E^{\alpha\beta\gamma\delta} * v_\delta|_\gamma \;\; with \;\; n^{\alpha\beta} = t * \sigma^{\alpha\beta}} \tag{6.5.3}$$

is valid, with the internal forces $n^{\alpha\beta}$, the plate thickness t, and the symmetric, positive-definite elasticity tensor $E^{\alpha\beta\gamma\delta}$ in the plane strain or plain stress state, including the elastic modulus E and the Poisson's ratio ν according to Hooke's law.

According to the ***least squares principle***, the minimum of (6.5.1) is sought by the first variation of $\tilde{E}$

$$\begin{aligned} \delta\tilde{E} &= -\iint_A \delta\hat{v}_\alpha|_\beta * E^{\alpha\beta\gamma\delta} * (v_\delta|_\gamma - \hat{v}_\delta|_\gamma)\, dA \\ &= -\iint_A \delta\hat{v}_\alpha|_\beta * (n^{\alpha\beta} - \hat{n}^{\alpha\beta})\, dA \;=\; 0 \end{aligned} \tag{6.5.4}$$

as a ***necessary condition***, where $\delta\hat{v}_\alpha$, also known as a virtual displacement, represents any variation of $\hat{v}_\alpha$. A variation of v_α is to be excluded with $\delta v_\alpha = 0$ because the exact solution is unalterable.

For non-trivial solutions, the second variation of $\tilde{E}$ provides the ***sufficient condition***

$$\delta^2\tilde{E} = \iint_A \delta\hat{v}_\alpha|_\beta * E^{\alpha\beta\gamma\delta} * \delta\hat{v}_\delta|_\gamma\, dA \; > 0 \tag{6.5.5}$$

for the existence of the minimum.

Using the Gauss/Green integral theorem (6.3.1), equation (6.5.4) yields

$$\delta\tilde{E} = -\int_\Gamma \delta\hat{v}_\alpha * (n^{\alpha\beta} - \hat{n}^{\alpha\beta}) * n_\beta\, d\Gamma + \iint_A \delta\hat{v}_\alpha * (n^{\alpha\beta}|_\beta - \hat{n}^\beta|_\beta)\, dA \;=\; 0.$$

In terms of the exact solution, the boundary conditions (6.3.3) and the equilibrium conditions (6.3.6) are to be inserted into this equation. The result for the continuum of the plate is then the well-known ***principle of virtual displacements*** (6.3.5)

$$\boxed{\delta\tilde{E} = \int_{\Gamma_\sigma} \delta\hat{v}_\alpha * (\hat{n}^{\alpha\beta} * n_\beta - \overline{p}^\alpha)\, d\Gamma - \iint_A \delta\hat{v}_\alpha * (\hat{n}^{\alpha\beta}|_\beta + \overline{q}^\alpha)\, dA \;=\; 0}\,. \tag{6.5.6}$$

Based on this principle ultimately, the question arises of how large the residual amounts for the obtained approximate solution.
For this purpose, the necessary condition (6.5.4) is used again. As $\delta\hat{v}_\alpha$ is arbitrary, this equation also holds for the special case of the exact solution v_α

$$-\iint_A v_\alpha|_\beta * E^{\alpha\beta\gamma\delta} * (v_\delta|_\gamma - \hat{v}_\delta|_\gamma)\, dA \;=\; 0,$$

respectively

$$\iint_A v_\alpha|_\beta * E^{\alpha\beta\gamma\delta} * \hat{v}_\delta|_\gamma \, dA \;=\; \iint_A v_\alpha|_\beta * E^{\alpha\beta\gamma\delta} * v_\delta|_\gamma \, dA.$$

By inserting these relations into (6.5.1) and considering the material law (6.5.3), the minimum of the residual is obtained as the ***error norm between the exact and approximate solutions***

$$\begin{aligned}
&\tilde{E}^{min} = \Pi^{(min)} - \hat{\Pi}^{(min)} \geq 0 \textit{ of the error order } O(h^{2*(p+1-1)}) \\
&\Pi^{(min)} = 1/2 * \iint_A v_\alpha|_\beta * E^{\alpha\beta\gamma\delta} * v_\delta|_\gamma \, dA \\
&\qquad = 1/2 * \iint_A v_\alpha|_\beta * n^{\alpha\beta} \, dA \\
&\hat{\Pi}^{(min)} = 1/2 * \iint_A \hat{v}_\alpha|_\beta * E^{\alpha\beta\gamma\delta} * \hat{v}_\delta|_\gamma \, dA \\
&\qquad = 1/2 * \iint_A \hat{v}_\alpha|_\beta * \hat{n}^{\alpha\beta} \, dA
\end{aligned} \tag{6.5.7}$$

with the energy difference of the potentials of internal forces.

Thus, the potential energy of the exact solution forms an upper bound for the potential energy of the FEM approximate solution depending on the specified error order (cf. (2.3.13))!

From this proof, it is evident with (6.5.1) that the ***finite element method represents a least squares method***, by which the approximation errors between the strains resp. stresses of the unknown exact solution and the approximate solution are minimized in an integral sense. The measure of convergence is determined, according to 6.5.2, primarily by the polynomial order p of the Taylor series used as approximation, and less by the measure h of discretization.

6.6 Stiffness Relations for Selected Examples

In the following, the already illustrated approach, of Section 2.1 at the level of the equation system (2.3.7) and the stiffness relationship of the complete system (2.3.6), is taken up again

$$\delta A = \delta V_I * (K_{IK} * V_K - F_I) = 0$$

for further explanations at the level of individual elements.
The complete system can be assembled by ***summation of the virtual work*** of all participating finite elements:

$$\boxed{\delta A = \delta A^{(i)} + \delta A^{(a)} = \sum(\delta a^{(i)} + \delta a^{(a)}) = 0} \tag{6.6.1}$$

$$\text{with } \delta A^{(i)} = \delta V_I * K_{IK} * V_K = \sum_{all}^{elements} \delta a^{(i)} \text{ and } \delta A^{(a)} = -\delta V_I * F_I = \sum_{all}^{elements} \delta a^{(a)}.$$

For this assembly, it is necessary to first determine the ***stiffness relationship of the element*** on the local level with the stiffness matrix $\boldsymbol{k}$, displacement vector $\boldsymbol{v}$, and load vector $\boldsymbol{f}$

$$\boxed{\begin{aligned} \delta a^{(i)} &= \delta v_I * k_{IK} * v_K \\ \delta a^{(a)} &= -\delta v_I * f_I \end{aligned}} . \tag{6.6.2}$$

In the following, the typical approach of the index-oriented matrix calculus will be presented and explained using two different examples. The formulations primarily serve to establish the necessary fundamentals for the application of object-oriented matrix algorithms.

The differential geometry of the elements in matrix calculus has already been discussed in Section 6.4. Now, building on (6.1.1), the FEM approximation of the displacement field is described to capture the state of the deformed body under external loads using methods of numerical mechanics. The focus is on linear, elastically deformable systems with small displacements and rotations. In the examples, the induced displacements, strains, and internal forces are determined as tensor fields, and they are calculated at selected points $P(\theta^\alpha)$ or $P(\lambda^\alpha)$ of the undeformed body.

For the determination of the stiffness matrix $\mathbf{k}$, the virtual work of the internal forces in the transformed form, as given by (6.3.5), is required:

$$\boxed{\delta a^{(i)} = \iint_A \delta v_i|_\beta * g_k{}^\beta * n^{ik} \, dA} , \tag{6.6.3}$$

and the material law (6.5.3) in the transformed form:

$$\boxed{\mathrm{n}^{ik} = E^{iklm} * g_m{}^{\delta} * v_l|_{\delta}} \tag{6.6.4}$$

with the doubly symmetric ***elasticity tensor***

$$E^{iklm} = E^{kilm} = E^{ikml} = E^{lmik}.$$

The ***material constants for an elastical isotropic material*** with the elasticity modulus E and Poisson's ratio ν for a plate of thickness t in Cartesian coordinates are given by

for the plane strain state

$$E\overbrace{ik}^{I}\overbrace{lm}^{K} = E^{IK} = \frac{E*t}{1-\nu^2} * \begin{pmatrix} 1 & 0 & 0 & \nu \\ 0 & \frac{1-\nu}{2} & \frac{1-\nu}{2} & 0 \\ 0 & \frac{1-\nu}{2} & \frac{1-\nu}{2} & 0 \\ \nu & 0 & 0 & 1 \end{pmatrix}$$

for the plane stress state

$$E\overbrace{ik}^{I}\overbrace{lm}^{K} = E^{IK} = \frac{E*t}{(1+\nu)*(1-2\nu)} * \begin{pmatrix} 1-\nu & 0 & 0 & 1 \\ 0 & \frac{1-2\nu}{2} & \frac{1-2\nu}{2} & 0 \\ 0 & \frac{1-2\nu}{2} & \frac{1-2\nu}{2} & 0 \\ 1 & 0 & 0 & 1-\nu \end{pmatrix} . \tag{6.6.5}$$

By substituting the material law (6.6.4) into (6.6.3), the virtual work of the internal forces is obtained for the determination of the stiffness matrix ***k***

$$\boxed{\delta a^{(i)} = \iint_A \delta v_i|_{\beta} * g_k{}^{\beta} * E^{iklm} * g_m{}^{\delta} * v_l|_{\delta}\, dA} . \tag{6.6.6}$$

For the determination of the load vector **f**, the virtual work of the external forces in the form

$$\boxed{\delta a^{(a)} = -\int_{L_n} \delta v_i * \overline{p}^i\, dL - \iint_A \delta v_i * \overline{q}^i\, dA} \tag{6.6.7}$$

is additionally required.

6.6.1 Stiffness Relationship of the Quadrilateral Element

As the first example for the formulation of the stiffness relationship, the quadrilateral element from Section 6.4.1 is selected. For the displacement field v_i according to (6.1.1), the same shape functions can be used as those used for definition of the geometry:

$$\Omega_{(K)}(\theta^\alpha) = N_{(K)}(\theta^\alpha)$$

with $\alpha \in \mathbb{N}\ [0,1]$. In such a case, it is refered as an ***isoparametric element*** of the form

$$\mathrm{v}_i = v_{i(K)} * \Omega_{(K)}\ , \tag{6.6.8}$$

where $v_{i(K)}$ is the ***vector of displacements*** at the element nodes (K).

The ***partial derivatives of these shape functions*** can also be obtained from the geometry:

$$\mathrm{v}_i|_\beta = v_{i(K)} * \Omega_{(K)},_\beta\ , \tag{6.6.9}$$

with

$$\Omega_{(K)},_\beta = N_{(K)},_\beta\ .$$

By substituting into (6.6.6) and considering the area element (6.3.7), the ***stiffness matrix k of the quadrilateral element*** is obtained in integral form as

$$\boxed{\begin{array}{c} \delta a^{(i)} = \delta v_{i(K)} * k_{i(K)\,l(M)} * v_{l(M)} \\ with \\ k_{i(K)\,l(M)} = \iint_A \Omega_{(K)},_\beta * g_k{}^\beta * E^{iklm} * g_m{}^\delta * \Omega_{(M)},_\delta * \sqrt{g}\ d\theta^0\ d\theta^1 \end{array}}\ , \tag{6.6.10}$$

where the surface integral still needs to be evaluated.

As the ***external load on the quadrilateral element***, according to (6.6.7), ***the surface load*** $\overline{q}^i$ is to be considered. For its distribution, the bilinear approach (6.4.2) as for the geometry of the bilinear quadrilateral is chosen:

$$\Psi_{(L)}(\theta^\alpha) = N_{(L)}(\theta^\alpha)$$

$$\overline{q}^i = \overline{q}^i_{(L)} * \Psi_{(L)} \quad \text{with} \quad \begin{array}{l} \Psi_{(0)} = 1/4*(1+\theta^0)*(1-\theta^1) \\ \Psi_{(1)} = 1/4*(1-\theta^0)*(1+\theta^1) \\ \Psi_{(2)} = 1/4*(1-\theta^0)*(1-\theta^1) \\ \Psi_{(3)} = 1/4*(1+\theta^0)*(1+\theta^1) \end{array} \tag{6.6.11}$$

with the ***vector*** $\overline{q}^i_{(L)}$ ***of surface load*** at the four corner nodes (L).

By substituting into (6.6.7) and considering the area element (6.3.7), the ***load matrix*** $\overline{\mathbf{f}}$ ***of the quadrilateral element*** is obtained in integral form as

$$\boxed{\begin{array}{c} \boldsymbol{\delta a}^{(a)} = \delta v_{i\textcircled{K}} * \overline{f}_{i\textcircled{K}} \\ \textit{with} \\ \overline{f}_{i\textcircled{K}} = \iint_A \Omega_{\textcircled{K}} * \Psi_{\textcircled{L}} * \sqrt{g}\, d\theta^0\, d\theta^1 * \overline{q}^i_{\textcircled{L}} \end{array}} , \qquad (6.6.12)$$

where the surface integral still needs to be evaluated.

Once the unknown displacements $v_{l\textcircled{L}}$ on the level of the overall system are computed, the ***internal forces of the element*** can be determined at any point $G(\theta^\alpha)$ according to (6.6.4):

$$\boxed{\boldsymbol{n}^{ik}_{\textcircled{G}} = E^{iklm} * g_m{}^\delta * \Omega_{\textcircled{L}},_\delta(\theta^\alpha_{\textcircled{G}}) * v_{l\textcircled{L}}} . \qquad (6.6.13)$$

For this purpose, the Gaussian points of the element are particularly suitable, offering higher approximation accuracy, in comparison to the nodal point. Their coordinates $\theta^\alpha_{\textcircled{G}}$ and weights $H_{\textcircled{G}}$ are referred by [Zienkiewicz/Taylor 1989].

For bilinear and quadratic quadrilateral elements described in Section 6.4.1, it is generally not possible to analytically perform the required integration of the area integrals in (6.6.10) and (6.6.12) because the contravariant basis vectors and the area element can be determined only pointwise and not explicitly.

Therefore, numerical ***Gauss integration*** must be used for this purpose, which approach is briefly explained in the following. It is assumed here that the elasticity tensor E^{iklm} is constant within the element and can therefore be pulled out of the integral.
For the computation of the ***stiffness matrix of the quadrilateral element*** would be to evaluate numerically

$$\boxed{\begin{array}{l} \boldsymbol{k}_{i\textcircled{K}\, l\textcircled{M}} = E^{iklm} * M_{\textcircled{K}\textcircled{M}}{}^{km} \\ \textit{the expression} \\ M_{\textcircled{K}\textcircled{M}}{}^{km} = \sum_{G=1}^{\textit{Gauss points}} H_{\textcircled{G}} * f(\theta^\alpha_{\textcircled{G}}) \\ \textit{with the Gaussian weights } H_{\textcircled{G}} \textit{ for the matrix function} \\ f(\theta^\alpha) = \Omega_{\textcircled{K}},_\beta * g_k{}^\beta * \sqrt{g} * g_m{}^\delta * \Omega_{\textcircled{M}},_\delta \\ \textit{at each Gaussian point } G(\theta^\alpha) . \end{array}} \qquad (6.6.14)$$

The numerical computation of the ***load matrix of the quadrilateral element*** is done in a similar way

$$\overline{f}_{i(K)} = \overline{q}^i_{(L)} * L_{(K)(L)}$$

by evaluation of the expression

$$L_{(K)(L)} = \Sigma_{G=1}^{Gauss\ points} H_{(G)} * f(\theta^\alpha_{(G)})$$

with the Gaussian weights $H_{(G)}$ *for the matrix function* (6.6.15)

$$f(\theta^\alpha) = \Omega_{(K)} * \Psi_{(L)} * \sqrt{g}$$

at each Gaussian point $G(\theta^\alpha)$.

bilinear Parallelogram Element

For the parallelogram element according to Section 6.4.1, as a special case of the general quadrilateral, the displayed area integrals can be evaluated exactly, as the area element (6.4.8) and the contravariant basis vectors (6.4.10) are explicitly known. Therefore, (6.6.10) and (6.6.12) are slightly modified, as the contravariant basis vectors and the area element are independent of the coordinates θ^α in this case.

The ***stiffness matrix of the parallelogram element*** (cf. the matrix structure in Section 3.2, Table 3.1) is given by

$$\begin{aligned} &k_{i(K)}{}^{l}{}_{(M)} = E^{il\beta\delta} * \sqrt{g} * M_{(K)(M)}{}^{\beta\delta} \\ &\text{with } E^{il\beta\delta} = g_k{}^{\delta} * E^{iklm} * g_m{}^{\beta} \\ &\text{and } M_{(K)(M)}{}^{\beta\delta} = \iint_A \Omega_{(K)},_{\beta} * \Omega_{(M)},_{\delta}\, d\theta^0\, d\theta^1 \end{aligned} \tag{6.6.16}$$

with the integrated matrices

$$M_{(K)(M)}{}^{00} = 1/12 * \begin{pmatrix} 4 & -2 & -4 & 2 \\ -2 & 4 & 2 & -4 \\ -4 & 2 & 4 & -2 \\ 2 & -4 & -2 & 4 \end{pmatrix} \quad M_{(K)(M)}{}^{01} = 1/12 * \begin{pmatrix} -3 & 3 & 3 & -3 \\ 3 & -3 & -3 & 3 \\ -3 & 3 & 3 & -3 \\ 3 & -3 & -3 & 3 \end{pmatrix}$$

$$M_{(K)(M)}{}^{10} = 1/12 * \begin{pmatrix} -3 & 3 & -3 & 3 \\ 3 & -3 & 3 & -3 \\ 3 & -3 & 3 & -3 \\ -3 & 3 & -3 & 3 \end{pmatrix} \quad M_{(K)(M)}{}^{11} = 1/12 * \begin{pmatrix} 4 & -2 & 2 & -4 \\ -2 & 4 & -4 & 2 \\ 2 & -4 & 4 & -2 \\ -4 & 2 & -2 & 4 \end{pmatrix}$$

and for the ***load vector of the parallelogram element*** follows:

$$\boxed{\begin{array}{l} \overline{f}_{i\,(K)} = \overline{q}^{i}{}_{(L)} * \sqrt{g} * L_{(K)(L)} \\ with\ L_{(K)(L)} = \iint_A \Omega_{(K)} * \Psi_{(L)}\, d\theta^0\, d\theta^1 \end{array}}$$

with the integrated matrix

$$L_{(K)(L)} = 1/9 * \begin{pmatrix} 4 & 1 & 2 & 2 \\ 1 & 4 & 2 & 2 \\ 2 & 2 & 4 & 1 \\ 2 & 2 & 1 & 4 \end{pmatrix}.$$

The ***internal forces of the parallelogram element*** are determined at the nodes (G) of the element according to (6.6.13):

$$\boxed{\begin{array}{l} \boldsymbol{n}^{ik}{}_{(G)} = E^{ikl\delta} * N_{(L)}{}^{\delta}{}_{(G)} * v_{l\,(L)} \\ where\ E^{ikl\delta} = E^{iklm} * g_m{}^{\delta} \\ and\ N_{(L)}{}^{\delta}{}_{(G)} = \Omega_{(L)},_{\delta}(\theta^{\alpha}{}_{(G)}) \end{array}} \qquad (6.6.18)$$

with the matrices of the four corner nodes

$$N_{(K)}{}^{\delta}{}_{(0)} = 1/4 * \begin{pmatrix} 2 & -2 \\ 0 & 0 \\ -2 & 0 \\ 0 & 2 \end{pmatrix} \quad N_{(K)}{}^{\delta}{}_{(1)} = 1/4 * \begin{pmatrix} 0 & 0 \\ -2 & 2 \\ 0 & -2 \\ 2 & 0 \end{pmatrix}$$

$$N_{(K)}{}^{\delta}{}_{(2)} = 1/4 * \begin{pmatrix} 2 & 0 \\ 0 & 2 \\ -2 & -2 \\ 0 & 0 \end{pmatrix} \quad N_{(K)}{}^{\delta}{}_{(3)} = 1/4 * \begin{pmatrix} 0 & -2 \\ -2 & 0 \\ 0 & 0 \\ 2 & 2 \end{pmatrix}.$$

6.6.2 Stiffness Relationship of the Triangle Element

To establish the stiffness relation of the triangle element according to Section 6.4.2, the procedure, already used for the quadrilateral element in Section 6.6.1, is largely followed. However, one crucial difference is that in this case, the linearly dependent triangle coordinates $\lambda^0 + \lambda^1 + \lambda^2 = 1$ are used.

For the displacement field v_i according to (6.1.1), the same shape functions can be used as in defining the geometry:

$$\Omega_{\textcircled{K}}(\lambda^\alpha) = N_{\textcircled{K}}(\lambda^\alpha),$$

where $\alpha \in \mathbb{N}\,[0,2]$. Thus, ***isoparametric elements*** of the form

$$v_i = v_{i\textcircled{K}} * \Omega_{\textcircled{K}} \tag{6.6.19}$$

are formed with the ***vector*** $v_{i\textcircled{K}}$ ***of displacements*** at the element nodes $\textcircled{K}$.

Also, the ***partial derivatives of these shape functions*** can be obtained from the geometry:

$$v_i|_\alpha = v_{i\textcircled{K}} * \Omega_{\textcircled{K}},_\alpha \tag{6.6.20}$$

with

$$\Omega_{\textcircled{K}},_\alpha = N_{\textcircled{K}},_\alpha .$$

By substituting into (6.6.6) and considering the area element (6.3.7), the ***stiffness matrix k of the triangular element*** is obtained in integral form as

$$\boxed{\begin{array}{c} \delta a^{(i)} = \delta v_{i\textcircled{K}} * k_{i\textcircled{K}}{}^{l}{}_{\textcircled{M}} * v_{l\textcircled{M}} \\ with \\ k_{i\textcircled{K}}{}^{l}{}_{\textcircled{M}} = \iint_A \Omega_{\textcircled{K}},_\beta * g_k{}^\beta * E^{iklm} * g_m{}^\delta * \Omega_{\textcircled{M}},_\delta * dA \end{array}}\,, \tag{6.6.21}$$

where the surface integral still needs to be evaluated.

As a ***loading of the triangle element***, the ***surface load*** $\overline{q}^i$ can be considered here using (6.6.7). For its distribution, the linear approach (6.4.2) is chosen, similar to the geometry of the linear triangle:

$$\Psi_{\textcircled{K}}(\lambda^\alpha) = N_{\textcircled{K}}(\lambda^\alpha)$$

$$\overline{q}^i = \overline{q}^i{}_{\textcircled{K}} * \Psi_{\textcircled{K}} \quad with \quad \begin{array}{l} \Psi_{\textcircled{0}} = \lambda^0 \\ \Psi_{\textcircled{1}} = \lambda^1 \\ \Psi_{\textcircled{2}} = \lambda^2 \end{array} \tag{6.6.22}$$

with the ***vector*** $\overline{q}^i{}_{\textcircled{K}}$ ***of the surface load*** at the three corner nodes $\textcircled{K}$.

By substituting into (6.6.7) and considering the area element (6.3.7), the ***load matrix*** $\overline{f}$ ***of the triangle element*** in integral form is obtained as

$$\boxed{\begin{array}{c} \delta a^{(a)} = \delta v_{i\textcircled{K}} * \overline{f}_{i\textcircled{K}} \\ with \\ \overline{f}_{i\textcircled{K}} = \iint_A \Omega_{\textcircled{K}} * \Psi_{\textcircled{L}}\, dA * \overline{q}^{i}_{\textcircled{L}} \end{array}}, \qquad (6.6.23)$$

where the surface integral still needs to be evaluated.

Once the unknown displacements $v_{l\textcircled{M}}$ on the level of the overall system are calculated, the ***internal forces of the element*** can be determined at any point $G(\lambda^\alpha)$ using (6.6.4):

$$\boxed{n^{ik}_{\textcircled{G}} = E^{iklm} * g_m{}^{\delta} * \Omega_{\textcircled{M}},_{\delta}(\lambda^{\alpha}_{\textcircled{G}}) * v_{l\textcircled{M}}}\,. \qquad (6.6.24)$$

For this purpose, the Gaussian points of the element are particularly suitable, primarily due to their higher approximation accuracy in comparison to the nodal points. Their coordinates $\lambda^{\alpha}_{\textcircled{G}}$ and weights $H_{\textcircled{G}}$ are to gather from the reference [Zienkiewicz/Taylor 1989].

For the quadratic triangle element described in Section 6.4.2, it is generally not possible to perform the required integration of the surface integrals in (6.6.21) and (6.6.23) analytically. This is because the contravariant basis vectors and the area element cannot be evaluated explicitly, but only pointwise.

Therefore, numerical ***Gauss integration*** must be used again for this purpose. It is assumed here that the elasticity tensor E^{iklm} is constant within the element and can therefore be pulled out of the integral.
Numerical evaluation is required then in (6.6.21) for the calculation of the ***stiffness matrix of the triangle element***

$$\boxed{\begin{array}{l} k_{i\textcircled{K}\,l\textcircled{M}} = E^{iklm} * M_{\textcircled{K}\textcircled{M}}{}^{km} \\ \textit{by evaluation of expression} \\ M_{\textcircled{K}\textcircled{M}}{}^{km} = \sum_{G=1}^{Gauss\ points} H_{\textcircled{G}} * f(\lambda^{\alpha}_{\textcircled{G}}) \\ \textit{with the Gaussian weights } H_{\textcircled{G}} \textit{ for the matrix function} \\ f(\lambda^{\alpha}) = \Omega_{\textcircled{K}},_{\beta} * g_k{}^{\beta} * \sqrt{g} * g_m{}^{\delta} * \Omega_{\textcircled{M}},_{\delta} \\ \textit{at each Gaussian point } G(\lambda^{\alpha})\,. \end{array}} \qquad (6.6.25)$$

Numerical computation of the ***load matrix of the triangle element*** (6.6.23) is done in a similar manner

$$\bar{f}_{i\textcircled{K}} = \bar{q}^{i}_{\textcircled{L}} * L_{\textcircled{K}\textcircled{L}}$$

by evaluation of expression

$$L_{\textcircled{K}\textcircled{L}} = \sum_{G=1}^{Gauss\ points} H_{\textcircled{G}} * f(\lambda^{\alpha}_{\textcircled{G}})$$

with the Gaussian weights $H_{\textcircled{G}}$ *for the matrix function*

$$f(\lambda^{\alpha}) = \Omega_{\textcircled{K}} * \Psi_{\textcircled{L}} * \sqrt{g}$$

at each Gaussian point $G(\lambda^{\alpha})$. (6.6.26)

linear Triangle Element

For the linear triangle element as described in Section 6.4.2, the surface integrals outlined above can be analytically evaluated. Therefore, (6.6.21) and (6.6.23) are slightly modified here since the contravariant basis vectors and the area element are independent of the coordinates λ^{α} in this case.

The ***stiffness matrix of the linear triangle element*** (cf. the matrix structure in Section 3.2, Table 3.1) is given by

$$k_{i\textcircled{K}}{}^{l}{}_{\textcircled{M}} = E^{il\beta\delta} * \sqrt{g} * M_{\textcircled{K}\textcircled{M}}{}^{\beta\delta}$$

$$with\ E^{il\beta\delta} = g_{k}{}^{\delta} * E^{iklm} * g_{m}{}^{\beta}$$

$$and\ M_{\textcircled{K}\textcircled{M}}{}^{\beta\delta} = 1/\sqrt{g} * \iint_{A} \Omega_{\textcircled{K}},{}^{\beta} * \Omega_{\textcircled{M}},{}^{\delta}\, dA \qquad (6.6.27)$$

with the integrated matrices

$$M_{\textcircled{K}\textcircled{M}}{}^{00} = 1/2 * \begin{pmatrix} 1&0&0 \\ 0&0&0 \\ 0&0&0 \end{pmatrix} \quad M_{\textcircled{K}\textcircled{M}}{}^{01} = 1/2 * \begin{pmatrix} 0&1&0 \\ 0&0&0 \\ 0&0&0 \end{pmatrix} \quad M_{\textcircled{K}\textcircled{M}}{}^{02} = 1/2 * \begin{pmatrix} 0&0&1 \\ 0&0&0 \\ 0&0&0 \end{pmatrix}$$

$$M_{\textcircled{K}\textcircled{M}}{}^{10} = 1/2 * \begin{pmatrix} 0&0&0 \\ 1&0&0 \\ 0&0&0 \end{pmatrix} \quad M_{\textcircled{K}\textcircled{M}}{}^{11} = 1/2 * \begin{pmatrix} 0&0&0 \\ 0&1&0 \\ 0&0&0 \end{pmatrix} \quad M_{\textcircled{K}\textcircled{M}}{}^{12} = 1/2 * \begin{pmatrix} 0&0&0 \\ 0&0&1 \\ 0&0&0 \end{pmatrix}$$

$$M_{\textcircled{K}\textcircled{M}}{}^{20} = 1/2 * \begin{pmatrix} 0&0&0 \\ 0&0&0 \\ 1&0&0 \end{pmatrix} \quad M_{\textcircled{K}\textcircled{M}}{}^{21} = 1/2 * \begin{pmatrix} 0&0&0 \\ 0&0&0 \\ 0&1&0 \end{pmatrix} \quad M_{\textcircled{K}\textcircled{M}}{}^{22} = 1/2 * \begin{pmatrix} 0&0&0 \\ 0&0&0 \\ 0&0&1 \end{pmatrix},$$

and for the ***load matrix of the linear triangle element*** follows:

$$\begin{array}{|l|}\hline \overline{f}_{i(K)} = \overline{q}^i_{(L)} * L_{(K)(L)} \\ with\ L_{(K)(L)} = \sqrt{g} * \iint_A \Omega_{(K)} * \Psi_{(L)}\, dA \\ \hline \end{array} \qquad (6.6.28)$$

with the integrated matrix

$$L_{(K)(L)} = \sqrt{g} * \begin{pmatrix} 1/2 & 0 & 0 \\ 0 & 1/2 & 0 \\ 0 & 0 & 1/2 \end{pmatrix}.$$

The ***internal forces of the linear triangle element*** can be calculated according to (6.6.24). However, it's important to note that, as per (6.4.15), the derivatives of the shape functions are constant in this case due to the δ -matrix $\delta_{(K)\delta}$:

$$\begin{array}{|l|}\hline \boldsymbol{n}^{ik} = E^{ikl\delta} * \delta_{(M)\delta} * v_{l(M)} \\ \quad = E^{ikl\delta} * v_{l(\delta)} \\ where\ E^{ikl\delta} = E^{iklm} * g_m{}^{\delta} \\ \hline \end{array} \quad . \qquad (6.6.29)$$

Due to the linear approximation for the displacement field, the internal forces in this element are therefore constant across the element.

References

Baker 1991. Baker, A. J.; Pepper, D. W.: Finite Elements 1-2-3. McGraw-Hill, New York (1991)

Bathe 1986. Bathe, K.-J.: Finite-Elemente-Methoden. Springer, Berlin (1986)

Clough/Penzien 1975. Clough, R. W.; Penzien, J.: Dynamics of Structures. McGraw-Hill, New York (1975)

Klingbeil 1966. Klingbeil, E.: Tensorrechnung für Ingenieure. Bibliographisches Institut, Mannheim (1966)

Meissner 1996. Meißner, U.: Ein übergeordnetes Variationsprinzip für den Fehlerabgleich der Finite-Element-Methode. Technische Mechanik 16, S. 333 (1996)

Meissner/Maurial 2000. Meißner, U. F.; Maurial, A.: Die Methode der Finiten Elemente - Eine Einführung in die Grundlagen. Springer, Berlin (2000)

Zienkiewicz/Taylor 1989. Zienkiewicz, O. C.; Taylor, R. L.: The Finite Element Method. McGraw-Hill, London (1989)

Chapter 7
Object-oriented Numerics for Finite Elements

Abstract

Finally, the application of object-oriented matrix calculus is presented in two sections about quadrilateral and triangular plate elements based on typical C++ routines. The numeric functions PARALLELOGRAM and TRIANGLE exemplify, in the source code, the calculations of differential geometries, stiffness matrix relationships, and internal forces for both finite elements. The computation of corresponding numerical results is also demonstrated. The elements discussed are a bilinear parallelogram element and a linear triangular element, each with abstracted elasticity parameters. This allows the verification of calculated results for deformations and stresses under tension and shear in an elementary way. The program code, kept simple and clear, primarily aims to demonstrate how the required matrix objects of the matrix classes can be created and managed. It also illustrates how tensor/matrix arithmetic can be handled intuitively and clearly with the overloaded syntax for matrix operations.

Following Chapter 6 with FEM theory of plate elements in tensor/matrix notation, the concrete implementation in object-oriented programming with C++ is demonstrated in the following using two typical examples. The examples are annotated with relevant comments and are largely self-explanatory. They demonstrate the advantageous, illustrative use of the new matrix objects and clarify the syntax of overloaded operators.

The matrices used consist, on the one hand, of rectangular matrices and vectors for rational figures of the classes MATRIX and VEKTOR, and on the other hand, for integers of the classes intMATRIX and intVEKTOR.
It is shown exemplarily how these matrices can be created at run time using constructors, either as static objects with round brackets or as dynamic objects with the pointer operator *'new'*. Dynamic matrices can be removed from memory using the operator *'delete'*, if necessary.

It should be noted that, as declared upon and as customary in FORTRAN, all new matrix objects are stored column-wise here. For the preallocation of matrices, the possibility has been created to use standard C-arrays with square brackets. However, as per the convention of C and C++, these arrays are stored row-wise. Therefore, in this special case, preallocation must be done in transposed form with rows and columns swapped.

U. F. Meissner, *Tensor Calculus with Object-Oriented Matrices for Numerical Methods in Mechanics and Engineering*, https://doi.org/10.1007/978-3-031-59302-4_7

The examples also demonstrate especially the extended use of overloaded operators from the C++ programming language, as documented and discussed in Chapter 3. These operators include both the usual arithmetic operators and the operators for standard input and output, as well as the C++ assignment operator '=' for objects, which corresponds to the mathematical assignment symbol ':='. Of course, understanding the use of these language elements is intuitively easy to follow.

A special advantage of object-oriented programming is the provided protection mechanisms against destruction and misuse of internal class structures. For example, the use of member functions is strictly bound to the respective class instances and can only be called using the dot or arrow operator in conjunction with an associated object as instance of the class. Outside the class, only the functions declared as public are available to the programmer for common use. In addition to private functions, the internal data and properties of class members are also hidden and protected. Access to them is exclusively via the declared public interfaces of the provided functions, following this encapsulation principle. These protection mechanisms, initially unfamiliar from the perspective of procedural programming languages, have proven to be a global standard by now.

For bundling up of indices, as described in Section 3.1, the member function *'dimension'* is available here. It is used in a way that the dimensions of a matrix object are temporarily modified for the execution of the respective operation and then reset back to their original state. This function is particularly important for the execution of the general matrix multiplication according to (2.1.10). The arrangement of indices in the multiplicands must be carefully considered because their order also determines the sequence in the resultant matrix. If the desired result in the desired order of indices cannot be achieved with multi-dimensional hypermatrices, it may be reached, as shown in the examples, by subsequently transposing of the resultant matrix.

As typical examples of numerical calculations, the following two sections each address, according to Section 6.6, the parallelogram element and the triangle element with the same dimensions and properties to keep the results comparable. Geometry, stiffness relationships, and internal forces of elementary deformation states of these elements are calculated to illustrate the properties of matrix objects and to deepen the ease of handling matrix operations.

7.1 Matrix Objects of the Element Function PARALLELOGRAM[1]

This section includes the computational function PARALLELOGRAM to demonstrate the specific features and advantages of the new matrix classes through a concrete example. The handling of its output stream outFILE by the associated main program FEM_Demo [Meissner 2023] was already outlined in Section 5.4.
The tensor/matrix calculus prepared in Sections 6.4.1 and 6.6.1 is now implemented.

The following quantities are determined in detail
for geometry:
- the basis vectors, their metrics, and the area element,
for stiffness relationships:
- the elasticity tensor, the stiffness matrix, and the load vector, as well as
- the force vector and the vector of reaction forces,
and for the inner stress of the plate:
- the internal forces
of the bilinear parallelogram element.

The individual program steps correspond to the mentioned theoretical fundamentals in their sequence and can be traced step by step with the inserted comments.

In the following passage, the object-oriented program code of the function PARALLELOGRAM in the C++ programming language is listed for the

Finite bilinear Parallelogram Element:

Listing 7.1: Function PARALLELOGRAM: Tensor/Matrix Algorithms of Finite Element

```
// ===================================================================
// Function PARALLELOGRAM for Demonstrating the Matrix Classes
//          -------------
// C++ - Matrix Objects and Methods for Tensor/Matrix Calculus
// ===================================================================
// Copyright: Univ.-Prof. Dr.-Ing. Udo F. Meissner, Erkrath/Darmstadt
// Advanced teaching and study program - Commercial use not permitted.
// The author assumes no liability for incompleteness and errors!
// Version:  2023
// Compiler: C++
// ===================================================================
// FEM: bilinear Parallelogram Element for elastic Plates:
// -----------------------------------------------------------
// Geometry, Elasticity Stiffness, Loads and internal Forces
// =============== Include Files =====================================
#include "pch.h"                                // pre-compiled header file
// -------------
// #include "Matrizen.h"
```

[1] Function PARALLELOGRAM provides some comments about original member function calls used in C++ source code of classes available on the WWW [Meissner 2023].

```
          // abstract Matrices class                       MATRIZEN
// #include "Matrix.h"
          // 4-dimensional double rectangular matrices     MATRIX
// #include "Vektor.h"
          // 1-dimensional vectors                         VEKTOR
// #include "delta_Matrix.h"
          // delta-matrices                                delta_MATRIX
// #include "e_Matrix.h"
          // e-matrices                                    e_MATRIX
// #include "intMatrix.h"
          // 4-dimensional integer rectangular matrices intMATRIX
// #include "intVektor.h"
          // 1-dimensional integer vectors                 intVEKTOR
// =====================================================================
//  PARALLELOGRAMM
int PARALLELOGRAM (ofstream& outFILE, ifstream& inFILE)
{                                          // outFILE logs the data input
                                           // inFile provides the data input
    // -----------------------------------------------------------------
    // ofstream outFILE;
    // string Output = "Output File";
    // outFILE.open(Output, ios::app);
    // cout << "Output File: " << Output << endl << endl;
    // =================================================================
    bool ERROR;
    ERROR = false;
    // *** ERROR = true;
    if (ERROR) std::cout << "Error␣=␣" << ERROR << endl;
    // ---------------------------------------------------
    // Dimension of Cartesian coordinates x(i)
    int dim_i = 2;
    // Dimension of natural coordinates T(alpha)
    int dim_n = 2;
    // Number of element nodes Node(P)
    int dim_P = 4;
    // =================================================================
    // Determine geometry with static matrices
    // =================================================================
    // External node numbers P in internal order
    // ------------------------------------------
    int i, branch;
    // Preset external node numbers P
    // ------------------------------
    int P4[4] = { 0, 1, 2, 4 };
    intVEKTOR P(P4, 4);
    //
    std::cout << endl;
    std::cout << "External␣node␣numbers␣P␣in␣internal␣order:"
                                                      << P << endl;
                                                outFILE << "" << endl;
    //
    // Natural coordinates theta(alpha,P) of element nodes P
    // (C array row-wise !!!)
    // -----------------------------------------------------------------
```

```
double P_theta[8] = {  1.0, -1.0,
                      -1.0,  1.0,
                      -1.0, -1.0,
                       1.0,  1.0 };
MATRIX theta_P(P_theta, dim_n, dim_P, 1);
std::cout
    << "Natural␣coordinates␣theta(alpha,P)␣of␣nodes␣P:"
                                          << theta_P << endl;
                                        outFILE << "" << endl;
//
// Coordinates x(i,P) of element nodes P
// #####################################
// Default
VEKTOR x_i0(dim_i), x_i1(dim_i), x_i2(dim_i), x_i3(dim_i);
//
  x_i0(0) = 6.0;  x_i1(0) =  1.0;  x_i2(0) = 1.0; x_i3(0) =  6.0;
  x_i0(1) = 2.0;  x_i1(1) = 12.0;  x_i2(1) = 2.0; x_i3(1) = 12.0;
//
// Test -> (5 * 5) with 45 degrees
// ------------------------------
//x_i0(0) = 6.0;  x_i1(0) =  6.0;  x_i2(0) = 1.0; x_i3(0) = 11.0;
//x_i0(1) = 2.0;  x_i1(1) =  7.0;  x_i2(1) = 2.0; x_i3(1) =  7.0;
//
// Test -> (5 * 10) with 90 degrees
// -------------------------------
//x_i0(0) = 6.0;  x_i1(0) =  1.0;  x_i2(0) = 1.0; x_i3(0) =  6.0;
//x_i0(1) = 2.0;  x_i1(1) = 12.0;  x_i2(1) = 2.0; x_i3(1) = 12.0;
//
// Test -> (5 * 10) with 63.4349 degrees
// ------------------------------------
//x_i0(0) = 1.0;  x_i1(0) =  0.5;  x_i2(0) = 0.0; x_i3(0) =  1.5;
//x_i0(1) = 0.0;  x_i1(1) =  1.0;  x_i2(1) = 0.0; x_i3(1) =  1.0;
// ----------------------------------------------------------------
// Input
std::cout << "Enter␣coordinates␣x(0)=x,␣x(1)=y␣of␣nodes␣P"<<endl;
std::cout << "->␣1␣=␣yes␣/␣0␣=␣no␣(default):␣" << endl;
cin >> i;                                  outFILE << i << endl;
if (i != 0)
{   // Enter nodes individually
    std::cout << "Coordinates␣of␣node␣" << P(0) << ":" << endl;
    cin >> x_i0;                        outFILE << x_i0 << endl;
    std::cout << "Coordinates␣of␣node␣" << P(1) << ":" << endl;
    cin >> x_i1;                        outFILE << x_i1 << endl;
    std::cout << "Coordinates␣of␣node␣" << P(2) << ":" << endl;
    cin >> x_i2;                        outFILE << x_i2 << endl;
    std::cout << "Coordinates␣of␣node␣" << P(3) << ":" << endl;
    cin >> x_i3;                        outFILE << x_i3 << endl;
};
// Store coordinates of all nodes x(i,P)
// -----_------------------------------
MATRIX x_iP(dim_i, dim_P);
x_iP = x_i0;
// x_iP.teilmatrix_hinein(0, 1) = x_i1;
   x_iP.partialmatrix_in (0, 1) = x_i1;
```

```
// x_iP.teilmatrix_hinein(0, 2) = x_i2;
   x_iP.partialmatrix_in (0, 2) = x_i2;
// x_iP.teilmatrix_hinein(0, 3) = x_i3;
   x_iP.partialmatrix_in (0, 3) = x_i3;
//
std::cout << endl;
std::cout << "Coordinates␣x(i,p)␣of␣nodes␣P:␣" << x_iP << endl;
                                                 outFILE << "" << endl;
// Geometry
// ---------
std::cout << "Differential␣Geometry" << endl;
std::cout << "=====================" << endl << endl;
//
// Determine                      g_cov(i, 0) = [x(i,0) - x(i,2)]/2,
// covariant basis vectors        g_cov(i, 1) = [x(i,1) - x(i,2)]/2
// ------------------------------------------------------------------
VEKTOR g_cov_0(dim_i), g_cov_1(dim_i);
g_cov_0 = (x_i0 - x_i2) * 0.5;      g_cov_1 = (x_i1 - x_i2) * 0.5;
//
MATRIX g_cov(dim_i, dim_n);
g_cov = g_cov_0;
// g_cov.teilmatrix_hinein(0, 1) = g_cov_1;
   g_cov.partialmatrix_in (0, 1) = g_cov_1;
//
std::cout << "Covariant␣basis␣vectors:␣" << g_cov << endl;
                                                 outFILE << "" << endl;
//
// Determine covariant metric
//                      Metric_cov = g_cov(i, apha) * g_cov(i, beta)
// ------------------------------------------------------------------
MATRIX Metric_cov(dim_n, dim_n);
Metric_cov =
    g_cov.dimension(1, dim_i, dim_n) *
                                    g_cov.dimension(1, dim_i, dim_n);
//
std::cout << "Covariant␣metric:␣" << Metric_cov << endl;
                                                 outFILE << "" << endl;
//
// Generate e-matrix
// -----------------
e_MATRIX e(dim_i, dim_i);
std::cout << "e-Matrix:␣" << e << endl;
                                                 outFILE << "" << endl;
//
// Calculate area element
//                A = det(g_cov) = e(I,K) * g_cov(I,0) * g_ckov(K,1)
// -------------------------------------------------------------------
double A;
MATRIX H(dim_i, dim_i);
H = e.dimension(1, dim_i, dim_i) * g_cov_0;
A = H.dimension(1, dim_i) * g_cov_1;
std::cout << "Area␣element␣=␣" << A << endl << endl;
//
//
```

```
// By means of tensorbasis transformation
// ========================================
// Determine contravariant basis vectors
// g_cont(I,alpha) = e(I,K) * g_cov(K,gamma) * e(alpha,gamma) / A
// -----------------------------------------------------------------
MATRIX g_cont(dim_n, dim_n);
// eps-Tensor
// ----------
e_MATRIX eps_cont(dim_i, dim_i, 1.0 / A);
g_cont = e * g_cov.dimension(1, dim_i, dim_n);
g_cont = g_cont * eps_cont;
// =================================================================
  // Alternativly: Determine contravariant basis vectors g_cont
  //               by Cholesky inversion of covariant metric
  // ===============================================================
    // MATRIX& Metric_cont = *new MATRIX(dim_n, dim_n);
    // Metric_cov.invertieren (Metric_cont, 1);
    // Metric_cov.invert      (Metric_cont, 1);
    // Transform g_cov by Metric_cont
    // g_cont = g_cov * Metric_cont;
    // delete& Metric_cont;
// =================================================================
std::cout << "Contravariant␣basis␣vectors:␣" << g_cont << endl;
                                              outFILE << "" << endl;
// -----------------------------------------------------------------
// Test g_cov(I, alpha) * g_cont(I, beta) = delta(alpha, beta)
H = g_cov.dimension(1, dim_n, dim_n) *
                                  g_cont.dimension(1,dim_n, dim_n);
std::cout << "Test:␣Orthogonality␣of␣basis␣vectors␣g:␣"
                               << H.dimension(dim_n, dim_n) << endl;
                                              outFILE << "" << endl;
double factor = A / 12.0;
//    // ===========================================================
 // Generate stiffness matrix with dynamic matrices
 // ================================================================
 //
 // Input plate stiffness E_xy in Cartesian coordinates
 // ################################################################
MATRIX E_xy(dim_i * dim_i, dim_i * dim_i);
// Young's modulus * plate thickness / Poisson's ratio nue
// Preallocate elastic constants
double t;
double E = 1.0;  double nue = 0.0;
// Input
std::cout << "Enter:␣>␣0␣thickness␣of␣the␣plate␣/" << endl;
std::cout << "␣␣␣␣␣␣␣=␣0␣␣(default:␣shear␣only)␣/␣" << endl;
std::cout << "␣␣␣␣␣␣␣<␣0␣␣(default:␣nue␣=␣0):␣" << endl;
cin >> t;                                     outFILE << t << endl;
//
if (t <= 0.0)
{
    if (t < 0.0) t = 1.0;
    E = 1.0;
    nue = 0.0;
```

```
        //
        branch = 1;
    }
    else
    {
        std::cout << "Enter␣Young's␣modulus:␣" << endl;
        cin >> E;                                outFILE << E << endl;
        std::cout << "Enter␣Poisson's␣ratio:␣" << endl;
        cin >> nue;                            outFILE << nue << endl;
        if (nue < 0.0)  nue = 0.0;
        if (nue >= 0.5) nue = 0.0;
        //
        // Input plate state
        // #################
        std::cout << "branch␣=␣1␣plane␣strain␣state␣/␣" << endl;
        std::cout << "␣␣␣␣␣␣␣=␣2␣plane␣stress␣state␣" << endl;
        std::cout << "Enter␣branch:␣";
        cin >> branch;                     outFILE << branch << endl;
    }
    // ---------------------------------------------------------------
    std::cout << endl;
    std::cout << "Plate␣thickness␣t:␣" << t << endl;
    std::cout << "Young's␣modulus␣E:␣" << E << endl;
    std::cout << "Poisson's␣ratio␣nue:␣" << nue << endl;
    std::cout << endl;
    // ---------------------------------------------------------------
      switch (branch)
    // ===============
    {
    case 1:
        // Plane strain state
        // ------------------
        std::cout << "Plane␣Strain␣State" << endl;
        std::cout << "==================" << endl << endl;
        //
        E = E * t / (1 - nue * nue);
        //
        E_xy(0, 0) = E;                E_xy(0, 3) = nue * E;
        E_xy(3, 0) = E_xy(0, 3);       E_xy(3, 3) = E_xy(0, 0);
        //
        if (t == 0.0) E = 1.0 / (1 - nue * nue);
        //
        E_xy(1, 1) = E_xy(1, 2) = E * (1 - nue) / 2;
        E_xy(2, 1) = E_xy(2, 2) = E_xy(1, 1);
        break;
        //
    case 2:
        // Plane stress state
        // ------------------
        std::cout << "Plane␣Stress␣State" << endl;
        std::cout << "==================" << endl << endl;
        //
        E = E * t / ((1 + nue) * (1 - 2 * nue * nue));
        //
```

```
        E_xy(0, 0) = E * (1 - nue);         E_xy(0, 3) = E;
        E_xy(1, 1) = E_xy(1, 2) = E * (1 - 2 * nue) / 2;
        E_xy(2, 1) = E_xy(2, 2) = E_xy(1, 1);
        E_xy(3, 0) = E_xy(0, 3);            E_xy(3, 3) = E_xy(0, 0);
        break;
    };
    // -----------------------------------------------------------------
    std::cout << "Elasticity␣matrix␣E(i,k,m,l):␣" << E_xy << endl;
                                                  outFILE << "" << endl;
    //
    // Elasticity Tensor E_contra(i,l,m,beta)
    //                            = E_xy(i,k,l,m) * g_contra(k,beta)
    // -----------------------------------------------------------------
    MATRIX& E_contra = *new MATRIX(dim_i, dim_i, dim_n, dim_n);
    //
    E_contra = E_xy.dimension(dim_i, dim_i, dim_i * dim_i) *
               g_cont.dimension(1, dim_i, dim_n);
    if (ERROR) std::cout
        << "Transformed␣elasticity␣tensor␣E_contra(i,k,l,delta):␣"
        << E_contra << endl;
    //
    // Elasticity tensor E_contra(i,l,beta,delta)
    //                        = E_xy(i,l,m,beta) * g_cont(m,delta)
    // -----------------------------------------------------------------
    E_contra = E_contra.dimension(dim_i * dim_i, dim_i, dim_n) *
               g_cont.dimension(1, dim_i, dim_n);
    //
    std::cout
        << "Transformed␣elasticity␣tensor␣E_contra(i,l,beta,delta):␣"
        << E_contra << endl;
                                                  outFILE << "" << endl;
    //
    std::cout
        << "Symmetric␣elasticity␣tensor␣E_contra(i,beta,l,delta):␣"
        //                          E_contra.transponieren
        << E_contra.transpose     (0, 2, 1, 3) << endl;
    //  << E_contra.transponieren(0, 2, 1, 3) << endl;
                                                  outFILE << "" << endl;
    //
    std::cout << "Determine␣Stiffness␣Matrix" << endl;
    std::cout << "==========================" << endl << endl;
    //
    // Preallocate integrated matrices for stiffness matrix
    // (C-arrays row-wise !!!)
    // -----------------------------------------------------------------
    factor = A / 12.0;
    //
    // Preallocation
    // -------------
    double Array_00[16] = { 4, -2, -4,  2,
                           -2,  4,  2, -4,
                           -4,  2,  4, -2,
                            2, -4, -2,  4 };
    MATRIX M_KM00 = *new MATRIX(Array_00, dim_P, dim_P);
```

```
//
if (ERROR)
    std::cout << "Matrix␣M(K,M,0,0):␣" << M_KM00 << endl;
//
// Preallocation
// -------------
double Array_10[16] = { -3,  3, -3,  3,
                         3, -3,  3, -3,
                         3, -3,  3, -3,
                        -3,  3, -3,  3 };
MATRIX& M_KM10 = *new MATRIX(Array_10, dim_P, dim_P);
// ------------------
if (ERROR)
    std::cout << "Matrix␣M(K,M,1,0):␣" << M_KM10 << endl;
//
// Preallocation
// -------------
double Array_01[16] = { -3,  3,  3, -3,
                         3, -3, -3,  3,
                        -3,  3,  3, -3,
                         3, -3, -3,  3 };
MATRIX& M_KM01 = *new MATRIX(Array_01, dim_P, dim_P);
// ------------------
if (ERROR)
    std::cout << "Matrix␣M(K,M,0,1):␣" << M_KM01 << endl;
//
// Preallocation
// -------------
double Array_11[16] = { 4, -2,  2, -4,
                       -2,  4, -4,  2,
                        2, -4,  4, -2,
                       -4,  2, -2,  4 };
MATRIX& M_KM11 = *new MATRIX(Array_11, dim_P, dim_P);
// ------------------
if (ERROR)
    std::cout << "Matrix␣M(K,M,1,1):␣" << M_KM11 << endl;
//
// Build dynamic matrix M(K, M, beta, delta)
// by storing the 4 individual matrices
// ------------------------------------------------------------------
MATRIX& M_KM_bd = *new MATRIX(dim_P * dim_P, dim_n, dim_n);
// -----------
// Storing M_K00
// M_KM_bd.teilmatrix_hinein(0) = M_KM00;
   M_KM_bd.partialmatrix_in (0) = M_KM00;
if (ERROR)
    std::cout << "Matrix␣M(KM_00):" << M_KM_bd << endl;
//
// Storing M_K10
// M_KM_bd.teilmatrix_hinein(0, 1, 0) = M_KM10;
   M_KM_bd.partialmatrix_in (0, 1, 0) = M_KM10;
if (ERROR)
    std::cout << "Matrix␣M(KM_10):" << M_KM_bd << endl;
//
```

```
// Storing M_KM01
// M_KM_bd.teilmatrix_hinein(0, 0, 1) = M_KM01;
   M_KM_bd.partialmatrix_in (0, 0, 1) = M_KM01;
if (ERROR)
    std::cout << "Matrix␣M(KM_01):" << M_KM_bd << endl;
//
//
// Storing M_KM11
// M_KM_bd.teilmatrix_hinein(0, 1, 1) = M_KM11;
   M_KM_bd.partialmatrix_in (0, 1, 1) = M_KM11;
if (ERROR)
    std::cout << "Matrix␣M(KM_11):" << M_KM_bd << endl;
//
//
std::cout << "Matrix␣M(KM,␣beta,␣delta)␣without␣Factor␣1/12:␣"
    << M_KM_bd.dimension(4, 4, 4) << endl;
                                            outFILE << "" << endl;
// Multiply M_KM_bd by A/12
M_KM_bd = M_KM_bd * factor;
if (ERROR)
   std::cout << "Matrix␣M(KM,␣beta,␣delta):␣" << M_KM_bd << endl;
//
// Delete submatrices
// ------------------
// ***delete& M_KM00;  delete& M_KM10;
// ***delete& M_KM01;  delete& M_KM11;
//
// Calculate auxiliary stiffness matrix Hsteif(i, l, K, M)
//   = E(i, l, beta, delta) * M_KM_bd(KM, beta, delta) <- incl. A
// -------------------------------------------------------------
// Create dynamic auxiliary matrix H_k
MATRIX& H_k = *new MATRIX(dim_i, dim_i, dim_P, dim_P);
//
H_k = E_contra.dimension(dim_i * dim_i, dim_n * dim_n) *
                  M_KM_bd.dimension(dim_P * dim_P, dim_n * dim_n);
if (ERROR)
    std::cout << "Matrix␣H_k(i,␣l,␣K,␣M):" << H_k << endl;
//
// Delete M_KM_bd, E_kontra
// ------------------------
delete& M_KM_bd;
delete& E_contra;
//
// =============================================================
// Calculate stiffness matrix k(i, K, l, M) = Hsteif(i, l, K, M)
// by transposing Hsteif
// =============================================================
MATRIX& k = *new MATRIX(dim_i, dim_P, dim_i, dim_P);
//
// k = H_k.transponieren(0, 2, 1, 3);
   k = H_k.transpose    (0, 2, 1, 3);
std::cout << "Stiffness␣matrix␣k(i,␣K,␣l,␣M):" << k << endl;
                                            outFILE << "" << endl;
//
```

```
// Delete auxiliary matrix H_k
// ------------------------
delete& H_k;
//
// ==============================================================
// Build load vector f_load(i, K) from surface load q(i, L)
// ==============================================================
//
std::cout << "Distributed␣Loads" << endl;
std::cout << "=================" << endl << endl;
//
// Input the load matrix q(i, L) at element nodes L
// ################################################
// Default values (C array row-wise !!!)
// ---------------------------------------
double Array_q1[8] = { 1, 1, 1, 1, 1, 1, 1, 1 };
// Test -> uniform load  = {1, 1, 1, 1, 1, 1, 1, 1};
//
MATRIX& q_iL = *new MATRIX(Array_q1, dim_i, dim_P);
//
VEKTOR q_i0(dim_i), q_i1(dim_i), q_i2(dim_i), q_i3(dim_i);
//
// Input
std::cout
    << "Enter␣surface␣loads␣q(0)=q(x),␣q(1)=q(y)␣of␣nodes␣L"
    << endl;
std::cout << "->␣1␣=␣yes␣/␣0␣=␣(null)␣/␣<␣0␣(uniform␣load␣q(i)):"
    << endl;
cin >> i;                                outFILE << i << endl;
//
if (i == 0)
{   // unloaded
    q_iL = q_iL * 0.0;
}
else
{
    if (i < 0)
    { // Uniform load
        std::cout << "Enter␣uniform␣load␣q(i)␣for␣all␣nodes:␣"
            << endl << endl;
        cin >> q_i0;                     outFILE << q_i0 << endl;
        q_i1 = q_i0;   q_i2 = q_i0;  q_i3 = q_i0;
    }
    if (i > 0)
    { // Input nodes individually
        std::cout
        << "Enter␣surface␣loads␣q(i)␣of␣all␣nodes␣individually:␣"
        << endl << endl;
        std::cout << "q(i)␣Node␣" << P(0) << ":" << endl;
        cin >> q_i0;                        outFILE << q_i0 << endl;
        std::cout << "q(i)␣Node␣" << P(1) << ":" << endl;
        cin >> q_i1;                        outFILE << q_i1 << endl;
        std::cout << "q(i)␣Node␣" << P(2) << ":" << endl;
        cin >> q_i2;                        outFILE << q_i2 << endl;
```

```
        std::cout << "q(i)␣Node␣" << P(3) << ":" << endl;
        cin >> q_i3;                                outFILE << q_i3 << endl;
    };
    // Store loads of all nodes q(i, P)
    q_iL = q_i0;
    // q_iL.teilmatrix_hinein(0, 1) = q_i1;
       q_iL.partialmatrix_in (0, 1) = q_i1;
    // q_iL.teilmatrix_hinein(0, 2) = q_i2;
       q_iL.partialmatrix_in (0, 2) = q_i2;
    // q_iL.teilmatrix_hinein(0, 3) = q_i3;
       q_iL.partialmatrix_in (0, 3) = q_i3;
};
//
std::cout << endl << "Surface␣loads␣q(i,␣L)␣of␣nodes␣L:"
          << q_iL << endl;
                                                outFILE << "" << endl;
//
// Preallocate integrated matrix N(K, L) for load vector
// -------------------------------------------------------
factor = A / 9.0;
//
// Default values (C array row-wise !!!)
// ---------------------------------------
double Array_KL[16] = { 4, 1, 2, 2,
                        1, 4, 2, 2,
                        2, 2, 4, 1,
                        2, 2, 1, 4 };
MATRIX& L_KL = *new MATRIX(Array_KL, dim_P, dim_P);
// ------------------
if (ERROR) std::cout << "Matrix␣L(K,␣L):␣" << L_KL << endl;
//
L_KL = L_KL * factor;
// ----------------
// Calculate load vector f_load(i, K) = q(i, L) * N(K, L) * A
// -----------------------------------------------------------
VEKTOR& f_load = *new VEKTOR(dim_i * dim_P);
//
f_load.dimension(dim_i, dim_P) = q_iL * L_KL;
// ------------------------------------------
std::cout
<< "Load␣vector␣f_load(i,␣K):␣" << f_load.dimension(dim_i, dim_P)
    << endl;
                                                outFILE << "" << endl;
//
// Delete uniform load matrices
// ----------------------------
delete& q_iL;  delete& L_KL;
//
// ================================================================
// Determine displacement states
// ================================================================
//
std::cout << "Displacement␣State" << endl;
std::cout << "==================" << endl << endl;
```

```
   //
   // Preallocate displacement vector (C array row-wise !!!)
   // ---------------------------------------------------------
   double Array_v[8] = { 0, 0, 0, 0, 0, 0, 0, 0 };
   VEKTOR v(Array_v, dim_i * dim_P);
   //
   // Test -> Shear x       = {0, 0, 1, 0, 0, 0, 1, 0};
   // Test -> Shear y       = {0, 1, 0, 0, 0, 0, 0, 1};
   // Test -> Tension x     = {1, 0, 0, 1, 0, 0, 1, 0};
   // Test -> Tension y     = {0, 0, 0, 1, 0, 0, 0, 1};
   // Test -> Rigid Body    = {1, 1, 1, 1, 1, 1, 1, 1};
   //
   // Input Displacement Vector
   // ##########################
   // Input
   std::cout
     << "Enter␣displacements␣v(0)=v(x),␣v(1)=v(y)␣of␣nodes" << endl;
   std::cout << "->␣1␣=␣yes␣/␣0␣=␣(null)␣/␣<␣0␣(Test):␣" << endl;
   std::cout <<
     "␣/-1␣(x-Tension)␣/-2␣(y-Tension)␣/-3␣(x-Shear)␣/-4␣(y-Shear):"
               << endl;
   cin >> i;                                 outFILE << i << endl;
   //
 if (i > 0)
 {
 std::cout << "Enter␣displacements␣v(1)␣of␣all␣nodes␣individually:␣"
           << endl << endl;
 std::cout << "v(1)␣Node␣" << P(0) << ":" << endl;     cin >> q_i0;
                                            outFILE << q_i0 << endl;
 std::cout << "v(1)␣Node␣" << P(1) << ":" << endl;     cin >> q_i1;
                                            outFILE << q_i1 << endl;
 std::cout << "v(1)␣Node␣" << P(2) << ":" << endl;     cin >> q_i2;
                                            outFILE << q_i2 << endl;
 std::cout << "v(1)␣Node␣" << P(3) << ":" << endl;     cin >> q_i3;
                                            outFILE << q_i3 << endl;
     //
     v = q_i0;
     // v.teilmatrix_hinein(2) = q_i1;
        v.partialmatrix_in (2) = q_i1;
     // v.teilmatrix_hinein(4) = q_i2;
        v.partialmatrix_in (4) = q_i2;
     // v.teilmatrix_hinein(6) = q_i3;
        v.partialmatrix_in (6) = q_i3;
   }
   else
   {
       if (i == 0)
       {
           std::cout << "Displacement␣vector␣v(1,␣M)␣=␣0␣" << endl;
       }
       else
       {
        i = i + 5;
        switch (i)
```

```
        {
        case 1:
          std::cout << "Displacement␣vector:␣Shear␣in␣y-direction"
                    << endl;
               v(1) = 1;    v(7) = 1;
               break;
        case 2:
          std::cout << "Displacement␣vector:␣Shear␣in␣x-direction"
                    << endl;
               v(2) = 1;    v(6) = 1;
               break;
        case 3:
          std::cout << "Displacement␣vector:␣Tension␣in␣y-direction"
                    << endl;
               v(3) = 1;    v(7) = 1;
               break;
         case 4:
          std::cout << "Displacement␣vector:␣Tension␣in␣x-direction"
                    << endl;
               v(0) = 1;    v(6) = 1;
               break;
         };
       };
  };
std::cout << endl << "Displacement␣vector␣v(l,␣M)␣of␣the␣nodes␣M:␣"
          << v.dimension(dim_i, dim_P) << endl;
                                                outFILE << "" << endl;
  //
  // Calculate force vector f(i, K) = k(i, K, l, M) * v(l, M)
  // ------------------------------------------------------------
  VEKTOR& f = *new VEKTOR(dim_i * dim_P);
  //
  f = k.dimension(dim_i * dim_P, dim_i * dim_P, 1) * v;
  //
  std::cout << "Force␣vector␣f(i,␣K)␣=␣k(i,␣k,␣l,␣m)␣*␣v(l,␣m):␣"
      << f.dimension(dim_i, dim_P) << endl;
                                               outFILE << "" << endl;
  //
  // Calculate reaction vector
  //              a(i, K) =  k(i, K, l, M) * v(l, M) - f_last(i, K)
  // ------------------------------------------------------------------
  VEKTOR& a = *new VEKTOR(dim_i * dim_P);
  //
  a = f - f_load;
  //
  std::cout << "Reaction␣vector␣a(i,␣K)␣=␣k(i,␣k,␣l,␣m)␣*"
            << "␣v(l,␣m)␣-␣f_load(i,␣K):␣"
            << a.dimension(dim_i, dim_P) << endl;
                                                outFILE << "" << endl;
  //
  // Delete force vectors f and a, stiffness matrix k
  // ---------------------------------------------------
  delete& f; delete& f_load;  delete& a;
  delete& k;
```

```
// ===================================================================
// Determine Internal Forces N(ik) at Nodes G
// ===================================================================
//
std::cout << "Internal␣Forces" << endl;
std::cout << "===============" << endl << endl;
//
// Preallocate integrated matrix N(K, delta) for internal forces
// ------------------------------------------------------------------
factor = 1.0 / 4.0;
//
// Preallocate (C arrays row-wise !!!)
// ------------------------------
double Array_0[8] = { 2, 0, -2, 0,
                     -2, 0,  0, 2 };
MATRIX N_Kd0(Array_0, dim_P, dim_n);
// --------------------------------
if (ERROR) std::cout << "Matrix␣N_Kd0(K,delta):␣"<< N_Kd0 <<endl;
//
N_Kd0 = N_Kd0 * factor;

//
double Array_1[8] = { 0, -2,  0, 2,
                      0,  2, -2, 0 };
MATRIX N_Kd1(Array_1, dim_P, dim_n);
// --------------------------------
if (ERROR) std::cout << "Matrix␣N(Kd_1):␣" << N_Kd1 << endl;
//
N_Kd1 = N_Kd1 * factor;
//
double Array_2[8] = { 2, 0, -2, 0,
                      0, 2, -2, 0 };
MATRIX N_Kd2(Array_2, dim_P, dim_n);
// --------------------------------
if (ERROR) std::cout << "Matrix␣N(Kd_2):␣" << N_Kd2 << endl;
//
N_Kd2 = N_Kd2 * factor;
//
double Array_3[8] = { 0, -2, 0, 2,
                     -2,  0, 0, 2 };
MATRIX N_Kd3(Array_3, dim_P, dim_n);
// --------------------------------
if (ERROR)  std::cout << "Matrix␣N(Kd_3):" << N_Kd3 << endl;
//
N_Kd3 = N_Kd3 * factor;
// ------------------------------------------------
// Build dynamic matrix N(K, delta, G)
//                    by incorporating the 4 individual matrices
// ------------------------------------------------------------------
MATRIX& N_KdG = *new MATRIX(dim_P, dim_n, dim_P);
// ------------------------------------------------
// Incorporate N_Kd0
// N_KdG.teilmatrix_hinein(0) = N_Kd0;
   N_KdG.partialmatrix_in (0) = N_Kd0;
```

```
if (ERROR) std::cout << "Matrix␣N(Kd_0):" << N_KdG << endl;
//
// Incorporate N_Kd1
// N_KdG.teilmatrix_hinein(0, 0, 1) = N_Kd1;
   N_KdG.partialmatrix_in (0, 0, 1) = N_Kd1;
if (ERROR) std::cout << "Matrix␣N(Kd_1):" << N_KdG << endl;
//
// Incorporate N_Kd2
// N_KdG.teilmatrix_hinein(0, 0, 2) = N_Kd2;
   N_KdG.partialmatrix_in (0, 0, 2) = N_Kd2;
if (ERROR) std::cout << "Matrix␣N(Kd_2):" << N_KdG << endl;
//
//
// Incorporate N_Kd3
// N_KdG.teilmatrix_hinein(0, 0, 3) = N_Kd3;
   N_KdG.partialmatrix_in (0, 0, 3) = N_Kd3;
if (ERROR) std::cout << "Matrix␣N(Kd_3):" << N_KdG << endl;
//
std::cout << "Matrix␣N(K,␣delta,␣G):␣" << N_KdG << endl;
                                          outFILE << "" << endl;
// ----------------------------------------------------------------
// Elasticity tensor E_cont(i, k, l, delta) =
//                             E_xy(i, k, l, m) * g_cont(m, delta)
// ----------------------------------------------------------------
MATRIX E_cont(dim_i, dim_i, dim_i, dim_n);
//
E_cont = E_xy.dimension(dim_i * dim_i * dim_i, dim_i) *
    g_cont.dimension(1, dim_i, dim_n);
//
std::cout
    << "Transformed␣elasticity␣tensor␣E_cont(i,␣k,␣l,␣delta):"
    << E_cont << endl;
                                          outFILE << "" << endl;
// ----------------------------------------------------------------
// Calculate internal forces n(i, K) at nodes G
// ----------------------------------------------
MATRIX n(dim_i, dim_i, dim_P);
// Calculate auxiliary matrix Hn(i, k, l, K, G) =
//                       E(i, k, l, delta) * N_KdG(K, delta, G)
// ----------------------------------------------------------------
// Generate dynamic auxiliary matrix Hn
MATRIX& Hn = *new MATRIX(dim_i * dim_i, dim_i, dim_P, dim_P);
//
Hn = E_cont.dimension(dim_i * dim_i * dim_i, dim_n) * N_KdG;
//
if (ERROR)
    std::cout << "Matrix␣Hn␣=␣E(i,␣k,␣l,␣K,␣G):␣" << Hn << endl;
// ---------------------------------------------------------------
// n(i, k, G) = Hn(i, k, l, K, G) * v(l, K)
//                             calculated with displacement vector
//
n = Hn.dimension(dim_i * dim_i, dim_i * dim_P, dim_P) *
    v.dimension(1, dim_i * dim_P);
//
```

```
    std::cout
        << "Internal␣forces␣n(i,␣K)␣at␣nodes␣G:␣" << n << endl;
                                                outFILE << "" << endl;
    // =================================================================
    std::cout << "End␣Parallelogram" << endl;
    std::cout << "=================" << endl;
    //
    std::cout << "Pause:" << endl;
    // ################################
    char c;
    cin >> c;                                      outFILE << c << endl;
    // =================================================================
    return 3;
};
// ======================================================================
```

As a numerical example for the parallelogram element, a rectangle with dimensions of width 5 and height 10 with a base angle of 90 degrees is chosen as illustrated in Fig. 7.1. This is to demonstrate comparable results for the differential geometry. As a simplification for the transparent interpretation of the calculation results, hypothetical values for the plate thickness $t = 1.0$, elasticity modulus $E = 1.0$, and Poisson's ratio $\nu = 0.0$ are assumed.

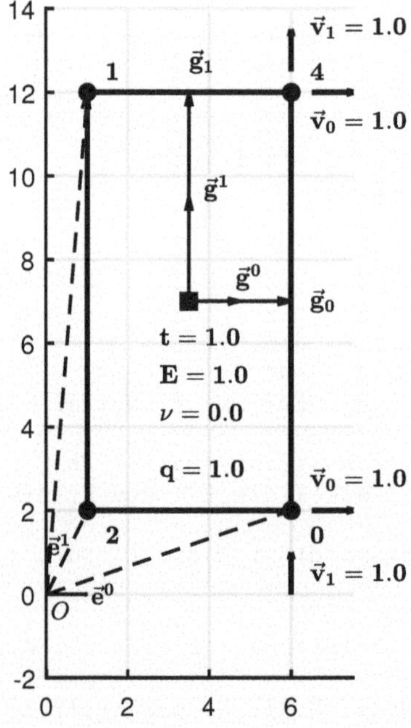

Fig. 7.1: Demonstration example of a bilinear parallelogram element under tension and shear, with an enlargement of the contravariant basis vectors $\mathbf{g}^\alpha$ by a factor of A/8

The calculation below was performed with a prescribed displacement field simulating pure tensile stress in the $\mathbf{e}^0$-direction to make the results easily verifiable. The specification of a uniform surface load is purely theoretical, aiming to illustrate the conversion into nodal loads of the load vector.
The results of the calculations are presented below:

Listing 7.2: Element Matrices of the PARALLELOGRAM Element with Tension in x-Direction

```
Finite bilinear Parallelogram Element
=====================================

Deformation in the x-direction due to Tension with given displacements
======================================================================

External node numbers P in internal order:
[ (1*4) * (1*1) ]-intMATRIX:
Columns 0 to 3
         0           1           2           4
Pause -> Enter a character:

Natural coordinates theta(alpha,P) of nodes P:
[ (2*4) * (1*1) ]-MATRIX:
Columns 0 to 3
  1.00e+00   -1.00e+00   -1.00e+00    1.00e+00
 -1.00e+00    1.00e+00   -1.00e+00    1.00e+00

Pause -> Enter a character:

Enter coordinates x(0)=x, x(1)=y of nodes P
-> 1 = yes / 0 = no (default): 0

Coordinates x(i,p) of nodes P:
[ (2*4) * (1*1) ]-MATRIX:
Columns 0 to 3
  6.00e+00    1.00e+00    1.00e+00    6.00e+00
  2.00e+00    1.20e+01    2.00e+00    1.20e+01

Pause -> Enter a character:

Differential Geometry
=====================

Covariant basis vectors:
[ (2*2) * (1*1) ]-MATRIX:
Columns 0 to 1
  2.50e+00    0.00e+00
  0.00e+00    5.00e+00

Pause -> Enter a character:

Covariant metric:
```

```
[ (2*2) * (1*1) ]-MATRIX:
Columns 0 to 1
  6.25e+00    0.00e+00
  0.00e+00    2.50e+01

Pause -> Enter a character:

e-Matrix:
[ (2*2) * (1*1) ]-MATRIX:
Columns 0 to 1
  0.00e+00    1.00e+00
 -1.00e+00    0.00e+00

Pause -> Enter a character:

Area element = 12.5

Contravariant basis vectors:
[ (2*2) * (1*1) ]-MATRIX:
Columns 0 to 1
  4.00e-01    0.00e+00
  0.00e+00    2.00e-01

Pause -> Enter a character:

Test: Orthogonality of basis vectors g:
[ (2*2) * (1*1) ]-MATRIX:
Columns 0 to 1
 1.000e+00  0.000e+00
 0.000e+00  1.000e+00

Pause -> Enter a character:

Enter: > 0 Thickness of the disc /
       = 0 (Default: shear only) /
       < 0 (Default: nue=0): -1

Plate thickness t: 1
Young's␣modulus␣E:␣1
Poisson's ratio nue: 0

Plane Strain State
==================

Elasticity matrix E(i,k,m,l):
[ (4*4) * (1*1) ]-MATRIX:
Columns 0 to 3
  1.00e+00    0.00e+00    0.00e+00    0.00e+00
  0.00e+00    5.00e-01    5.00e-01    0.00e+00
  0.00e+00    5.00e-01    5.00e-01    0.00e+00
  0.00e+00    0.00e+00    0.00e+00    1.00e+00

Pause -> Enter a character:
```

```
Transformed elasticity tensor E_contra(i,l,beta,delta):
[ (2*2) * (2*2) ]-MATRIX:
Columns 0 to 3
  1.60e-01   0.00e+00   0.00e+00   2.00e-02
  0.00e+00   0.00e+00   4.00e-02   0.00e+00
  0.00e+00   4.00e-02   0.00e+00   0.00e+00
  8.00e-02   0.00e+00   0.00e+00   4.00e-02

Pause -> Enter a character:

Symmetric elasticity tensor E_contra(i,beta,l,delta):
[ (2*2) * (2*2) ]-MATRIX:
Columns 0 to 3
  1.60e-01   0.00e+00   0.00e+00   0.00e+00
  0.00e+00   8.00e-02   4.00e-02   0.00e+00
  0.00e+00   4.00e-02   2.00e-02   0.00e+00
  0.00e+00   0.00e+00   0.00e+00   4.00e-02

Pause -> Enter a character:

Matrix M(KM, beta, delta) without factor 1/12:
[ (4*4) * (4*1) ]-MATRIX:
Columns 0 to 3
  4.00e+00  -3.00e+00  -3.00e+00   4.00e+00
 -2.00e+00   3.00e+00   3.00e+00  -2.00e+00
 -4.00e+00  -3.00e+00   3.00e+00   2.00e+00
  2.00e+00   3.00e+00  -3.00e+00  -4.00e+00
 -2.00e+00   3.00e+00   3.00e+00  -2.00e+00
  4.00e+00  -3.00e+00  -3.00e+00   4.00e+00
  2.00e+00   3.00e+00  -3.00e+00  -4.00e+00
 -4.00e+00  -3.00e+00   3.00e+00   2.00e+00
 -4.00e+00   3.00e+00  -3.00e+00   2.00e+00
  2.00e+00  -3.00e+00   3.00e+00  -4.00e+00
  4.00e+00   3.00e+00   3.00e+00   4.00e+00
 -2.00e+00  -3.00e+00  -3.00e+00  -2.00e+00
  2.00e+00  -3.00e+00   3.00e+00  -4.00e+00
 -4.00e+00   3.00e+00  -3.00e+00   2.00e+00
 -2.00e+00  -3.00e+00  -3.00e+00  -2.00e+00
  4.00e+00   3.00e+00   3.00e+00   4.00e+00

Pause -> Enter a character:

Stiffness matrix k(i,K,l,M):
[ (2*4) * (2*4) ]-MATRIX:
Columns 0 to 5
  7.50e-01  -1.25e-01  -3.75e-01   1.25e-01  -6.25e-01   1.25e-01
 -1.25e-01   5.00e-01   1.25e-01  -2.50e-01  -1.25e-01  -2.50e-01
 -3.75e-01   1.25e-01   7.50e-01  -1.25e-01   2.50e-01  -1.25e-01
  1.25e-01  -2.50e-01  -1.25e-01   5.00e-01   1.25e-01   0.00e+00
 -6.25e-01  -1.25e-01   2.50e-01   1.25e-01   7.50e-01   1.25e-01
  1.25e-01  -2.50e-01  -1.25e-01   0.00e+00   1.25e-01   5.00e-01
  2.50e-01   1.25e-01  -6.25e-01  -1.25e-01  -3.75e-01  -1.25e-01
 -1.25e-01   0.00e+00   1.25e-01  -2.50e-01  -1.25e-01  -2.50e-01
```

```
Columns 6 to 7
  2.50e-01   -1.25e-01
  1.25e-01    0.00e+00
 -6.25e-01    1.25e-01
 -1.25e-01   -2.50e-01
 -3.75e-01   -1.25e-01
 -1.25e-01   -2.50e-01
  7.50e-01    1.25e-01
  1.25e-01    5.00e-01

Pause -> Enter a character:

Distributed Loads
=================

Enter surface loads q(0)=q(x), q(1)=q(y) at nodes L
-> 1 = yes / 0 = (zero) / < 0 (Uniform load q(i)):
-1
Input of uniform load q(i) for all nodes:

 [ (1*2) * (1*1) ]-MATRIX (row-wise !!!) input:
Enter input:
1 1

Surface loads q(i,L) at nodes L:
[ (2*4) * (1*1) ]-MATRIX:
Columns 0 to 3
  1.00e+00    1.00e+00    1.00e+00    1.00e+00
  1.00e+00    1.00e+00    1.00e+00    1.00e+00

Pause -> Enter a character:

Load vector f_load(i,K):
[ (2*4) * (1*1) ]-MATRIX:
Columns 0 to 3
  1.25e+01    1.25e+01    1.25e+01    1.25e+01
  1.25e+01    1.25e+01    1.25e+01    1.25e+01

Pause -> Enter a character:

Displacement State
===================

Enter displacements v(0)=v(x), v(1)=v(y) of nodes
 -> 1 = ja / 0 = (zero) / < 0 (Test):
 / -1 (x-Tension) / -2 (y-Tension) /-3 (x-Shear) / -4 (y-Shear):
-1
Displacement vector: Tension in x-direction

Displacement vector v(l,M) of the nodes M:
[ (2*4) * (1*1) ]-MATRIX:
Columns 0 to 3
  1.00e+00    0.00e+00    0.00e+00    1.00e+00
  0.00e+00    0.00e+00    0.00e+00    0.00e+00
```

```
Pause -> Enter a character:

Force vector f(i,K) = k(i,k,l,m) * v(l,m):
[ (2*4) * (1*1) ]-MATRIX:
Columns 0 to 3
  1.00e+00  -1.00e+00  -1.00e+00   1.00e+00
  0.00e+00   0.00e+00   0.00e+00   0.00e+00

Pause -> Enter a character:

Reaction vector a(i,K) = k(i,k,l,m) * v(l,m) - f_load(i,K):
[ (2*4) * (1*1) ]-MATRIX:
Spalten 0 bis 3
 -1.15e+01  -1.35e+01  -1.35e+01  -1.15e+01
 -1.25e+01  -1.25e+01  -1.25e+01  -1.25e+01

Pause -> Zeichen eingeben:

Internal Forces
===============

Matrix N(K, delta, G):
[ (4*2) * (4*1) ]-MATRIX:
Spalten 0 bis 3
  5.00e-01   0.00e+00   5.00e-01   0.00e+00
  0.00e+00  -5.00e-01   0.00e+00  -5.00e-01
 -5.00e-01   0.00e+00  -5.00e-01   0.00e+00
  0.00e+00   5.00e-01   0.00e+00   5.00e-01
 -5.00e-01   0.00e+00   0.00e+00  -5.00e-01
  0.00e+00   5.00e-01   5.00e-01   0.00e+00
  0.00e+00  -5.00e-01  -5.00e-01   0.00e+00
  5.00e-01   0.00e+00   0.00e+00   5.00e-01

Pause -> Enter a character:

transformed Elasticity tensor E_kont(i,k,l,delta):
[ (2*2) * (2*2) ]-MATRIX:
Columns 0 to 3
  4.00e-01   0.00e+00   0.00e+00   0.00e+00
  0.00e+00   2.00e-01   1.00e-01   0.00e+00
  0.00e+00   2.00e-01   1.00e-01   0.00e+00
  0.00e+00   0.00e+00   0.00e+00   2.00e-01

Pause -> Enter a character:

Internal Forces
===============

Internal forces n(i,K) at nodes G:
[ (2*2) * (4*1) ]-MATRIX:
Columns 0 to 3
  2.00e-01   2.00e-01   2.00e-01   2.00e-01
  0.00e+00   0.00e+00   0.00e+00   0.00e+00
```

```
  0.00e+00    0.00e+00    0.00e+00    0.00e+00
  0.00e+00    0.00e+00    0.00e+00    0.00e+00

Pause -> Enter a character:

End Parallelogram
=================
```

To verify the results, a second calculation for the same system is performed with a modified displacement field to generate pure shear deformation in the $\mathbf{e}^1$-direction. The abbreviated listing below displays the results:

Listing 7.3: Element Matrices of the PARALLELOGRAM Element with Shear in y-Direction

```
Finite bilinear Parallelogram Element
=====================================

Deformation in the y-direction due to Shear with given displacements
====================================================================

External node numbers P in internal order:
[ (1*4) * (1*1) ]-intMATRIX:
Columns 0 to 3
         0           1           2           4
Pause -> Enter a character:

........

Generated Stiffness k(i,K,l,M):
[ (2*4) * (2*4) ]-MATRIX:
Columns 0 to 5
  7.50e-01  -1.25e-01  -3.75e-01   1.25e-01  -6.25e-01   1.25e-01
 -1.25e-01   5.00e-01   1.25e-01  -2.50e-01  -1.25e-01  -2.50e-01
 -3.75e-01   1.25e-01   7.50e-01  -1.25e-01   2.50e-01  -1.25e-01
  1.25e-01  -2.50e-01  -1.25e-01   5.00e-01   1.25e-01   0.00e+00
 -6.25e-01  -1.25e-01   2.50e-01   1.25e-01   7.50e-01   1.25e-01
  1.25e-01  -2.50e-01  -1.25e-01   0.00e+00   1.25e-01   5.00e-01
  2.50e-01   1.25e-01  -6.25e-01  -1.25e-01  -3.75e-01  -1.25e-01
 -1.25e-01   0.00e+00   1.25e-01  -2.50e-01  -1.25e-01  -2.50e-01

Columns 6 to 7
  2.50e-01  -1.25e-01
  1.25e-01   0.00e+00
 -6.25e-01   1.25e-01
 -1.25e-01  -2.50e-01
 -3.75e-01  -1.25e-01
 -1.25e-01  -2.50e-01
  7.50e-01   1.25e-01
  1.25e-01   5.00e-01

Pause -> Enter a character:

Enter displacements v(0)=v(x), v(1)=v(y) of nodes
 -> 1 = yes / 0 = (zero) / < 0 (Test):
```

```
 / -1 (x-Tension) / -2 (y-Tension) /-3 (x-Shear) / -4 (y-Shear):
-4
Displacement vector: Shear in y-direction

Displacement vector v(l,M) of the nodes M:
[ (2*4) * (1*1) ]-MATRIX:
Columns 0 to 3
  0.00e+00    0.00e+00    0.00e+00    0.00e+00
  1.00e+00    0.00e+00    0.00e+00    1.00e+00

Pause -> Enter a character:

Force vector f(i,K) = k(i,k,l,m) * v(l,m):
[ (2*4) * (1*1) ]-MATRIX:
Columns 0 to 3
 -2.50e-01    2.50e-01   -2.50e-01    2.50e-01
  5.00e-01   -5.00e-01   -5.00e-01    5.00e-01

Pause -> Enter a character:

Internal Forces
===============

Internal forces n(i,K) at nodes G:
[ (2*2) * (4*1) ]-MATRIX:
Columns 0 to 3
  0.00e+00    0.00e+00    0.00e+00    0.00e+00
  1.00e-01    1.00e-01    1.00e-01    1.00e-01
  1.00e-01    1.00e-01    1.00e-01    1.00e-01
  0.00e+00    0.00e+00    0.00e+00    0.00e+00

Pause -> Zeichen eingeben:

End Parallelogram
=================
```

The properties of the parallelogram element can be verified and examined based on the numerical results of both calculations, according to the elementary equilibrium and deformation checks of mechanics/statics, along with the unit displacement states contained in the stiffness matrix.

7.2 Matrix Objects of the Element Function TRIANGLE[2]

This section includes the computational function TRIANGLE to demonstrate the specific features and advantages of the new matrix classes through another example.
The handling of its output stream outFILE by the associated main program FEM_Demo [Meissner 2023] was already outlined in Section 5.4.
The tensor/matrix calculus prepared in Sections 6.4.2 and 6.6.2 is now implemented.

The following quantities are determined in detail
for geometry:
- the basis vectors, their metrics, and the area element,
for stiffness relationships:
- the elasticity tensor, the stiffness matrix, and the load vector, as well as
- the force vector and the vector of reaction forces,
and for the internal stress of the plate:
- the internal forces
of the linear triangular element.

The individual program steps correspond to the mentioned theoretical fundamentals in their sequence and can be traced step by step with the inserted comments.

The calculation of differential geometry for the triangular element has a different structure than for the parallelogram element, because it is based on the natural triangle coordinates (6.4.11), which are linearly dependent on each other and derived from the virtual tetrahedron.
In the following passage, the object-oriented program code of the function TRIANGLE in the C++ programming language is listed for the

Finite linear Triangle Element:

Listing 7.4: Function TRIANGLE: Tensor/Matrix Algorithms of Finite Element

```
// ================================================================
// Function TRIANGLE for Demonstrating the Matrix Classes
//          --------
// C++ - MATRIX Objects and Methods for Tensor/Matrix Calculus
// ================================================================
// Copyright: Univ.-Prof. Dr.-Ing. Udo F. Meissner, Erkrath/Darmstadt
// Advanced teaching and study program - Commercial use not permitted.
// The author assumes no liability for incompleteness and errors!
// Version:  2023
// Compiler: C++
// ================================================================
// FEM: linear Triangular Element for elastic Plates:
// -----------------------------------------------
```

[2] Function TRIANGLE provides some comments about original member function calls used in C++ source code of classes available on the WWW [Meissner 2023].

```
// Geometry, Elasticity, Stiffness, Loads and internal Forces
// =============== Include Files ======================================
#include "pch.h"                               // pre-compiled header file
// -------------
// #include "Matrizen.h"
            // abstract Matrices class                       MATRIZEN
// #include "Matrix.h"
            // 4-dimensional double rectangle matrices      MATRIX
// #include "Vektor.h"
            // 1-dimensional vectors                         VEKTOR
// #include "delta_Matrix.h"
            // delta-matrices                                delta_MATRIX
// #include "e_Matrix.h"
            // e-matrices                                    e_MATRIX
// #include "intMatrix.h"
            // 4-dimensional integer matrices                intMATRIX
// #include "intVektor.h"
            // 1-dimensional integer vectors                 intVEKTOR
// ====================================================================
//  DREIECK
int TRIANGLE(ofstream& outFILE, ifstream& inFILE)
{                          // outFILE logs the data input
                           // inFile provides the data input
    // ----------------------------------------------------------------
    // ofstream outFILE;
    // string output = "Output file";
    // outFILE.open(output, ios::app);
    // cout << "Output file: " << Output << endl << endl;
    // ================================================================
    bool ERROR;
    ERROR = false;
    // ERROR = true;
    if (ERROR) std::cout << "Error␣=␣" << ERROR << endl;
    // -------------------------------------------------------
    // Dimension of Cartesian coordinates x(i)
    int dim_i = 2;
    // Dimension of natural coordinates T(alpha)
    int dim_n = 3;
    // Number of element nodes Node(P)
    int dim_P = 3;
    // ================================================================
    // Determine geometry with static matrices
    // ================================================================
    // External node numbers P in internal order
    // ------------------------------------------
    int i, branch;
    // Preset external node numbers P
    // -----------------------------
    int P3[3] = { 0, 1, 2 };
    intVEKTOR P(P3, dim_P);
    //
    std::cout << endl;
    std::cout << "External␣node␣numbers␣P␣in␣internal␣order:"
                                                     << P << endl;
```

```
                                              outFILE << "" << endl;
//
// Natural coordinates theta(alpha,P) of element nodes P
// (C field row-wise !!!)
// -----------------------------------------------------------------
double P_theta[9] = { 1.0,  0.0, 0.0,
                      0.0,  1.0, 0.0,
                      0.0,  0.0, 1.0 };
MATRIX theta_P(P_theta, dim_n, dim_P, 1);
std::cout
      << "Natural␣coordinates␣theta(alpha,P)␣of␣nodes␣P:"
                                                << theta_P << endl;
                                              outFILE << "" << endl;
//
// Coordinates x(i,P) of element nodes P
// ####################################
// Default
VEKTOR x_i0(dim_i), x_i1(dim_i), x_i2(dim_i);
   x_i0(0) = 6.0;   x_i1(0) =  1.0;   x_i2(0) = 1.0;
   x_i0(1) = 2.0;   x_i1(1) = 12.0;   x_i2(1) = 2.0;
//
// Test -> (5 * 5) with 45 degrees
// --------------------------------
// x_i0(0) = 6.0;   x_i1(0) =  6.0;   x_i2(0) = 1.0;
// x_i0(1) = 2.0;   x_i1(1) =  7.0;   x_i2(1) = 2.0;
//
// Test -> (5 * 10) with 90 degrees
// --------------------------------
// x_i0(0) = 6.0;   x_i1(0) =  1.0;   x_i2(0) = 1.0;
// x_i0(1) = 2.0;   x_i1(1) = 12.0;   x_i2(1) = 2.0;
//
// Test -> (5 * 10) with 63.4349 degrees
// -------------------------------------
// x_i0(0) = 1.0;   x_i1(0) =  0.5;   x_i2(0) = 0.0;
// x_i0(1) = 0.0;   x_i1(1) =  1.0;   x_i2(1) = 0.0;
//
// ---------------------------------------------------
// Input
std::cout <<"Enter␣coordinates␣x(0)=x,␣x(1)=y␣of␣nodes␣P" <<endl;
std::cout <<"->␣1␣=␣yes␣/␣0␣=␣no␣(default):␣" << endl;
cin >> i;                                      outFILE << i << endl;
if (i != 0)
{   // Enter nodes individually
    std::cout << "Coordinates␣of␣node␣" << P(0) << ":" << endl;
    cin >> x_i0;                             outFILE << x_i0 << endl;
    std::cout << "Coordinates␣of␣node␣" << P(1) << ":" << endl;
    cin >> x_i1;                             outFILE << x_i1 << endl;
    std::cout << "Coordinates␣of␣node␣" << P(2) << ":" << endl;
    cin >> x_i2;                             outFILE << x_i2 << endl;
};
// Store coordinates of all nodes x(i,P)
// -----_------------------------------
MATRIX x_iP(dim_i, dim_P);
x_iP = x_i0;
```

```
 //
 // x_iP.teilmatrix_hinein(0, 1) = x_i1;
    x_iP.partialmatrix_in (0, 1) = x_i1;
 // x_iP.teilmatrix_hinein(0, 2) = x_i2;
    x_iP.partialmatrix_in (0, 2) = x_i2;
 //
 std::cout << endl;
 std::cout << "Coordinates␣x(i,p)␣of␣nodes␣P:␣" << x_iP << endl;
                                                   outFILE << "" << endl;
 // Geometry
 // ---------
 std::cout << "Differential␣Geometry" << endl;
 std::cout << "=====================" << endl << endl;
//
// Generate covariant basis vectors g_cov
// -------------------------------------------------------------------
MATRIX g_cov(dim_i, dim_n);
g_cov = x_iP;
//
std::cout << "Covariant␣basis␣vectors␣g:␣" << g_cov << endl;
                                                  outFILE << "" << endl;
// -------------------------------------------------------------------
// Determine covariant basis vectors a_cov(I, n)
// -------------------------------------------------------------------
VEKTOR a_cov_0(dim_n), a_cov_1(dim_n), a_cov_2(dim_n);
// Individual vectors
a_cov_0 = x_i0;   a_cov_0(2) = 1.0;
a_cov_1 = x_i1;   a_cov_1(2) = 1.0;
a_cov_2 = x_i2;   a_cov_2(2) = 1.0;
if (ERROR) std::cout << "a_cov_0␣=␣" << a_cov_0 << "a_cov_1␣=␣"
                << a_cov_1 << "a_cov_2␣=␣" << a_cov_2 << endl;
//
MATRIX a_cov(dim_n, dim_n);
a_cov = a_cov_0;
//
// a_cov.teilmatrix_hinein(0, 1) =
//                 a_cov_1; a_cov.teilmatrix_hinein(0, 2) = a_cov_2;
   a_cov.partialmatrix_in(0, 1) =
                   a_cov_1; a_cov.partialmatrix_in (0, 2) = a_cov_2;
//
std::cout
      << "Covariant␣basis␣vectors␣a␣of␣the␣virtual␣tetrahedron:␣"
                                                     << a_cov << endl;
                                                  outFILE << "" << endl;
//
// Determine covariant metric
//           Metric_cov = a_cov(i, alpha) * a_cov(i, beta)
// -------------------------------------------------------------------
MATRIX Metric_cov(dim_n, dim_n);
Metric_cov =
a_cov.dimension(1, dim_n, dim_n) *
                            a_cov.dimension(1, dim_n, dim_n);
//
std::cout
```

```
      << "Covariant␣metric␣of␣a_cov␣of␣the␣virtual␣tetrahedron:␣"
                                          << Metric_cov << endl;
                                          outFILE << "" << endl;
//
// Generate e-matrix
// -----------------
e_MATRIX e(dim_n, dim_n, dim_n);
std::cout << "e-Matrix:␣" << e << endl;
                                          outFILE << "" << endl;
//
// Calculate area element
// A = det(a_cov) =
//              e(I,K,L) * a_cov(I,0) * a_cov(K,1) * a_cov(L,2)
// ----------------------------------------------------------------
// Determine contravariant basis vectors a_cont(n, n)
// ----------------------------------------------------------------
double A;
MATRIX& H = *new MATRIX(dim_n, dim_n);
H = e.dimension(1, dim_n, dim_n * dim_n) * a_cov_0;
H = H.dimension(1, dim_n, dim_n) * a_cov_1;
A = H.dimension(1, dim_n) * a_cov_2;
std::cout << "Area␣element␣=␣" << A << endl << endl;
//
// By means of tensorbasis transformation
// ======================================
// Calculate contravariant basis vectors
// a_cont(I,beta) = 1/2 * e(I,L,N) * a_cov(L,delta) = H1 * H2 =
//                  a_cov(N, lambda) * e(delta, lambda, beta) / A
// ----------------------------------------------------------------
MATRIX a_cont(dim_n, dim_n);
MATRIX& H1 = *new MATRIX(dim_n, dim_n, dim_n);
MATRIX& H2 = *new MATRIX(dim_n, dim_n, dim_n);
// Form epsilon tensor
// --------------------
e_MATRIX eps_cont(dim_n, dim_n, dim_n, 1.0/A);
H1 = e * a_cov.dimension(1, dim_n, dim_n);
if (ERROR) std::cout << "H1␣=␣" << H1 << endl;
H2 = a_cov * eps_cont;
if (ERROR) std::cout << "H2␣=␣" << H2 << endl;
//
// a_cont(I,beta) = H1 * H2
a_cont = H1.dimension(dim_n, dim_n*dim_n,1) *
                              H2.dimension(1, dim_n*dim_n,dim_n);
// ----------------------------------------------------------------
if (ERROR) std::cout << "a_cont␣=␣" << a_cont << endl;
//
a_cont = 0.5 * a_cont;
// Delete H1, H2
delete& H1;
delete& H2;
// ================================================================
  // Alternatively: Determine contravariant basis vectors a_cont
  //        by Gauss inversion (nonsymmetric) of covariant metric
  // ==============================================================
```

```
   // MATRIX& H3 = *new MATRIX(dim_n, dim_n);
   //// H3 = a_cov.transponieren();
   //   H3 = a_cov.transpose    ();
   //// A = H3.invertieren(a_cont, -1);
   //   A = H3.invert      (a_cont, -1);
   // delete& H3;
// ==============================================================
std::cout
     << "Contravariant␣basis␣vectors␣of␣the␣virtual␣tetrahedron:"
                                               << a_cont << endl;
                                           outFILE << "" << endl;
// --------------------------------------------------------------
// Test: a_cov(I, alpha) * a_cont(I, beta) = delta(alpha, beta)
H=a_cov.dimension(1,dim_n,dim_n) *
                               a_cont.dimension(1, dim_n, dim_n);
std::cout << "Test:␣Orthogonality␣of␣the␣basis␣vectors␣a";
std::cout
   << "␣of␣the␣virtual␣tetrahedron:␣" << H.dimension(dim_n,dim_n)
                                                         << endl;
                                           outFILE << "" << endl;
// Contravariant basis vectors g_cont
// -----------------------------------
MATRIX g_cont(dim_i, dim_n);
g_cont(0,0) = a_cont(0,0);   g_cont(1,0) = a_cont(1,0);
g_cont(0,1) = a_cont(0,1);   g_cont(1,1) = a_cont(1,1);
g_cont(0,2) = a_cont(0,2);   g_cont(1,2) = a_cont(1,2);
std::cout << "Contravariant␣basis␣vectors␣g:␣" << g_cont << endl;
                                           outFILE << "" << endl;
// --------------------------------------------------------------
// Test: Sum g_cov = 0
H(0) = H(1) = 0.0;
for (i = 0; i < 3;  i++) { H(0) = H(0) + g_cont(0, i);
                             H(1) = H(1) + g_cont(1, i); }
std::cout << "Test:␣Sum␣of␣contravariant␣basis␣vectors␣g:␣"
                                 << H.dimension(dim_i) << endl;
                                           outFILE << "" << endl;
// --------------------------------------------------------------
// Test: g_cov(i, alpha) * g_cont(i, beta) = delta(alpha, beta)
H =g_cov.dimension(1, dim_i,dim_n) *
                                g_cont.dimension(1,dim_i,dim_n);
if (ERROR)
     std::cout<<"!!!␣No␣orthogonality␣of␣the␣basis␣vectors␣g␣!!!"
                           << H.dimension(dim_n, dim_n) << endl;
// ==============================================================
// Generate stiffness matrix with dynamic matrices
// ==============================================================
//
// Input plate stiffness E_xy in Cartesian coordinates
// ##################################################
MATRIX E_xy(dim_i * dim_i, dim_i * dim_i);
// Young's modulus * plate thickness / Poisson's ratio nue
// Preallocate elastic constants
double t;
double E = 1.0;  double nue = 0.0;
```

```
// Input
std::cout << "Enter:␣>␣0␣thickness␣of␣the␣plate␣/" << endl;
std::cout << "␣␣␣␣␣␣␣=␣0␣␣(default:␣shear␣only)␣/␣" << endl;
std::cout << "␣␣␣␣␣␣␣<␣0␣␣(default:␣nue␣=␣0):␣" << endl;
cin >> t;                                   outFILE << t << endl;
//
if (t <= 0.0)
{
   if (t < 0.0) t = 1.0;
   E = 1.0;
   nue = 0.0;
   //
   branch = 1;
}
else
{
   std::cout << "Enter␣Young's␣modulus:␣"<< endl;
   cin >> E;                                   outFILE << E << endl;
   std::cout << "Enter␣Poisson's␣ratio:␣"<< endl;
   cin >> nue;                               outFILE << nue << endl;
   if (nue < 0.0)  nue = 0.0;
   if (nue >= 0.5) nue = 0.0;
   //
   // Input plate state
   // #################
   std::cout << "branch␣=␣1␣plane␣strain␣state␣/␣" << endl;
   std::cout << "␣␣␣␣␣␣␣=␣2␣plane␣stress␣state␣" << endl;
   std::cout << "Enter␣branch:␣";
   cin >> branch;                         outFILE << branch << endl;
}
// ------------------------------------------------------------------
std::cout << endl;
std::cout << "Plate␣thickness␣t:␣" << t << endl;
std::cout << "Young's␣modulus␣E:␣" << E << endl;
std::cout << "Poisson's␣ratio␣nue:␣" << nue << endl;
std::cout << endl;
// ------------------------------------------------------------------
   switch (branch)
// ===============
{
case 1:
   // Plane strain state
   // ------------------
   std::cout << "Plane␣Strain␣State" << endl;
   std::cout << "==================" << endl << endl;
   //
   E = E * t / (1 - nue * nue);
   //
   E_xy(0, 0) = E;                    E_xy(0, 3) = nue * E;
   E_xy(3, 0) = E_xy(0, 3);           E_xy(3, 3) = E_xy(0, 0);
   //
   if (t == 0.0) E = 1.0 / (1 - nue * nue);
   //
   E_xy(1, 1) = E_xy(1, 2) = E * (1 - nue) / 2;
```

```
    E_xy(2, 1) = E_xy(2, 2) = E_xy(1, 1);
    break;
    //
  case 2:
    // Plane stress state
    // ------------------
    std::cout << "Plane␣Stress␣State" << endl;
    std::cout << "==================" << endl << endl;
    //
    E = E * t / ((1 + nue) * (1 - 2 * nue * nue));
    //
    E_xy(0, 0) = E * (1 - nue);          E_xy(0, 3) = E;
    E_xy(1, 1) = E_xy(1, 2) = E * (1 - 2 * nue) / 2;
    E_xy(2, 1) = E_xy(2, 2) = E_xy(1, 1);
    E_xy(3, 0) = E_xy(0, 3);             E_xy(3, 3) = E_xy(0, 0);
    break;
};
// ----------------------------------------------------------------
std::cout << "Elasticity␣matrix␣E(i,k,m,l):␣" << E_xy << endl;
                                            outFILE << "" << endl;
//
// Elasticity Tensor E_contra(i,l,m,beta)
//                    = E_xy(i,k,l,m) * g_contra(k,beta)
// ----------------------------------------------------------------
MATRIX& E_contra = *new MATRIX(dim_i, dim_i, dim_n, dim_n);
//
E_contra = E_xy.dimension(dim_i, dim_i, dim_i * dim_i) *
                           g_cont.dimension(1, dim_i, dim_n);
if (ERROR) std::cout
 << "Transformed␣elasticity␣tensor␣E_contra(i,k,l,delta):␣"
                                            << E_contra << endl;
//
// Elasticity tensor E_contra(i,l,beta,delta)
//                       = E_xy(i,l,m,beta) * g_cont(m,delta)
// ----------------------------------------------------------------
E_contra = E_contra.dimension(dim_i * dim_i, dim_i, dim_n)
                            * g_cont.dimension(1, dim_i, dim_n);
//
std::cout
    <<"Transformed␣elasticity␣tensor␣E_contra(i,l,beta,delta):␣"
                                             << E_contra << endl;
                                            outFILE << "" << endl;
//
std::cout
      <<"Symmetric␣elasticity␣tensor␣E_contra(i,beta,l,delta):␣"
                  << E_contra.transpose     (0, 2, 1, 3) << endl;
//                << E_contra.transponieren(0, 2, 1, 3) << endl;

                                            outFILE << "" << endl;
std::cout << "Determine␣Stiffness␣Matrix" << endl;
std::cout << "==========================" << endl << endl;
//
// Preallocate integrated matrices for stiffness matrix
// ----------------------------------------------------------------
```

```
MATRIX& M_KM_bd = *new MATRIX(dim_P * dim_P, dim_n * dim_n);
double factor;  factor = A / 2.0;
//
 // Preallocate dynamic matrix M(K,M,beta,delta)
 // -------------------------------------------
 // Values of 9 partial matrices M_KM_00 ..
 //                              M_KM_22 store directly
 M_KM_bd(0,0) = 1.0;   M_KM_bd(1,1) = 1.0;  M_KM_bd(2,2) = 1.0;
 M_KM_bd(3,3) = 1.0;   M_KM_bd(4,4) = 1.0;  M_KM_bd(5,5) = 1.0;
 M_KM_bd(6,6) = 1.0;   M_KM_bd(7,7) = 1.0;  M_KM_bd(8,8) = 1.0;
 //
 std::cout << "Matrix␣M(K,M,␣beta,delta)␣without␣Factor␣1/2:␣"
                                            << M_KM_bd << endl;
                                         outFILE << "" << endl;
 // M_KM_bd mit A/2 multiplizieren
 M_KM_bd = M_KM_bd * factor;
 if (ERROR) std::cout << "Matrix␣M(KM,␣beta,delta):␣"
                                            << M_KM_bd << endl;
 //
 // Calculate auxiliary stiffness matrix Hsteif(i,l,K,M) =
 //      E(i,l,beta,delta) * M_KM_bd(KM,beta,delta) <- incl. A
 // -------------------------------------------------------------
 // Create dynamic auxiliary matrix H_k anlegen
 MATRIX& H_k = *new MATRIX(dim_i, dim_i, dim_P, dim_P);
 //
 H_k = E_contra.dimension(dim_i * dim_i,dim_n * dim_n) *
                M_KM_bd.dimension(dim_P * dim_P, dim_n * dim_n);
 if (ERROR) std::cout << "Matrix␣H_k(i,l,K,M):" << H_k <<endl;
 //
 // Delete M_KM_bd, E_kontra
 // ------------------------
 delete& M_KM_bd;
 delete& E_contra;
 //
 // ==================================================================
 // Calculate stiffness matrix k(i,K,l,M) = Hsteif(i,l,K,M)
 // by transposing Hsteif
 // ==================================================================
 MATRIX&  k = *new MATRIX(dim_i, dim_P, dim_i, dim_P);
 //
 // k = H_k.transponieren(0, 2, 1, 3);
    k = H_k.transpose    (0, 2, 1, 3);
 std::cout << "Stiffness␣matrix␣k(i,K,l,M):" << k << endl;
                                         outFILE << "" << endl;
 // Delete auxiliary matrix H_k
 // ---------------------------
 delete& H_k;
 //
 // ==================================================================
 // Build load vector f_load(i, K) from surface load q(i, L)
 // ==================================================================
//
std::cout << "Distributed␣Loads" << endl;
std::cout << "=================" << endl << endl;
```

```
  //
  // Input the load matrix q(i, L) at element nodes L
  // ##############################################
  // Default values (C array row-wise !!!)
  // -----------------------------------
  double Array_q1[6] = {1, 1, 1, 1, 1, 1};
  // Test -> uniform load  = {1, 1, 1, 1, 1, 1};
  //
  MATRIX& q_iL = *new MATRIX(Array_q1, dim_i, dim_P);
  //
  VEKTOR q_i0(dim_i), q_i1(dim_i), q_i2(dim_i);
  //
 // Input
  std::cout
  << "Enter␣surface␣loads␣q(0)=q(x),␣q(1)=q(y)␣of␣nodes␣L"
                                                          << endl;
  std::cout << "->␣1␣=␣yes␣/␣0␣=␣(null)␣/␣<␣0␣(uniform␣load␣q(i)):"
                                                           << endl;
  cin >> i;                                   outFILE << i << endl;
  //
  if (i == 0)
  {   // unloaded
      q_iL = q_iL * 0.0;
  }
  else
  {
      if (i < 0)
      { // Uniform load
           std::cout << "Enter␣uniform␣load␣q(i)␣for␣all␣nodes:␣"
                                                  << endl << endl;
           cin >> q_i0;                       outFILE << q_i0 << endl;
           q_i1 = q_i0;    q_i2 = q_i0;
   }
   if (i > 0)
   { // Input nodes individually
        std::cout
            << "Enter␣surface␣loads␣q(i)␣of␣all␣nodes␣individually:␣"
            << endl << endl;
        std::cout << "q(i)␣Node␣" << P(0) << ":" << endl;
        cin >> q_i0;                          outFILE << q_i0 << endl;
        std::cout << "q(i)␣Node␣" << P(1) << ":" << endl;
        cin >> q_i1;                          outFILE << q_i1 << endl;
        std::cout << "q(i)␣Node␣" << P(2) << ":" << endl;
        cin >> q_i2;                          outFILE << q_i2 << endl;
   };
   // Store loads of all nodes q(i, P)
   q_iL = q_i0;
   // q_iL.teilmatrix_hinein(0, 1) = q_i1;
      q_iL.partialmatrix_in (0, 1) = q_i1;
   // q_iL.teilmatrix_hinein(0, 2) = q_i2;
      q_iL.partialmatrix_in (0, 2) = q_i2;
};
//
std::cout << endl << "Surface␣loads␣q(i,␣L)␣of␣nodes␣L:"
```

```
                    << q_iL << endl;
                                                    outFILE << "" << endl;
//
// Preallocate integrated matrix N(K, L) for load vector
// --------------------------------------------------------
factor = A / 24.0;
//
// Default values (C array row-wise !!!)
// ---------------------------------------
double Array_KL[9] = { 2, 1, 1,
                       1, 2, 1,
                       1, 1, 2};
MATRIX& L_KL = *new MATRIX(Array_KL, dim_P, dim_P);
// ------------------
if (ERROR) std::cout << "Matrix␣L(K,␣L):␣" << L_KL << endl;
//
L_KL = L_KL * factor;
// ------------------
// Calculate load vector f_load(i, K) = q(i, L) * N(K, L) * A
// -------------------------------------------------------------
VEKTOR& f_load = *new VEKTOR(dim_i * dim_P);
//
f_load.dimension(dim_i, dim_P) = q_iL * L_KL;
// ------------------------------------------
std::cout
    << "Load␣vector␣f_load(i,␣K):␣" << f_load.dimension(dim_i, dim_P)
    << endl;
                                                    outFILE << "" << endl;
//
// Delete uniform load matrices
// ----------------------------
delete& q_iL;  delete& L_KL;
//
// =====================================================================
// Determine displacement states
// =====================================================================
//
std::cout << "Displacement␣State" << endl;
std::cout << "==================" << endl << endl;
//
// Preallocate displacement vector (C array row-wise !!!)
// -------------------------------------------------------
double Array_v[6] = {0, 0, 0, 0, 0, 0};
VEKTOR v(Array_v, dim_i * dim_P);
//
// Test -> Tension x    = {1, 0, 0, 0, 0, 0};
// Test -> Tension y    = {0, 0, 0, 1, 0, 0};
// Test -> Shear x      = {0, 0, 1, 0, 0, 0};
// Test -> Shear y      = {0, 1, 0, 0, 0, 0};
// Test -> Rigid Body   = {1, 1, 1, 1, 1, 1};
//
// Input Displacement Vector
// #########################
// Input
```

```
std::cout
    << "Enter␣displacements␣v(0)=v(x),␣v(1)=v(y)␣of␣nodes" << endl;
std::cout << "->␣1␣=␣yes␣/␣0␣=␣(null)␣/␣<␣0␣(Test):␣" << endl;
std::cout <<
     "␣/-1␣(x-Tension)␣/-2␣(y-Tension)␣/-3␣(x-Shear)␣/-4␣(y-Shear):"
         << endl;
cin >> i;                                           outFILE << i << endl;
//
if (i > 0)
 {std::cout << "Enter␣displacements␣v(1)␣of␣all␣nodes␣individually:␣"
           << endl << endl;
    std::cout << "v(1)␣Node␣" << P(0) << ":" << endl;     cin >> q_i0;
                                                 outFILE << q_i0 << endl;
    std::cout << "v(1)␣Node␣" << P(1) << ":" << endl;     cin >> q_i1;
                                                 outFILE << q_i1 << endl;
    std::cout << "v(1)␣Node␣" << P(2) << ":" << endl;     cin >> q_i2;
                                                 outFILE << q_i2 << endl;
 //
    v = q_i0;
 // v.teilmatrix_hinein(2) = q_i1;
    v.partialmatrix_in (2) = q_i1;
 // v.teilmatrix_hinein(4) = q_i2;
    v.partialmatrix_in (4) = q_i2;
 }
else
 {
    if (i == 0)
    {
        std::cout << "Displacement␣vector␣v(1,␣M)␣=␣0␣" << endl;
    }
    else
    {
        i = i + 5;
  switch (i)
        {
  case 1:
          std::cout << "Displacement␣vector:␣Shear␣in␣y-direction"
                    << endl;
          v(1) = 1;
          break;
 case 2:
          std::cout << "Displacement␣vector:␣Shear␣in␣x-direction"
                    << endl;
          v(2) = 1;
          break;
 case 3:
          std::cout << "Displacement␣vector:␣Tension␣in␣y-direction"
                    << endl;
          v(3) = 1;
          break;
 case 4:
          std::cout << "Displacement␣vector:␣Tension␣in␣x-direction"
                    << endl;
          v(0) = 1;
```

```
          break;
         };
     };
 };
 std::cout << endl << "Displacement␣vector␣v(l,␣M)␣of␣the␣nodes␣M:␣"
          << v.dimension(dim_i, dim_P) << endl;
                                        outFILE << "" << endl;
//
// Calculate force vector f(i, K) = k(i, K, l, M) * v(l, M)
// ------------------------------------------------------------
VEKTOR& f = *new VEKTOR(dim_i * dim_P);
//
f = k.dimension(dim_i * dim_P, dim_i * dim_P, 1) * v;
//
std::cout << "Force␣vector␣f(i,␣K)␣=␣k(i,␣k,␣l,␣m)␣*␣v(l,␣m):␣"
         << f.dimension(dim_i, dim_P) << endl;
                                        outFILE << "" << endl;
//
// Calculate reaction vector
//                  a(i, K) =  k(i, K, l, M) * v(l, M) - f_last(i, K)
// ---------------------------------------------------------------------
VEKTOR& a = *new VEKTOR(dim_i * dim_P);
//
a = f - f_load;
//
std::cout <<
  "Reaction␣vector␣a(i,␣K)␣=␣k(i,␣k,␣l,␣m)␣*␣v(l,␣m)␣-␣f_load(i,␣K):"
         << a.dimension(dim_i, dim_P) << endl;
                                             outFILE << "" << endl;
//
// Delete force vectors f and a, stiffness matrix k
// ----------------------------------------------------
delete& f; delete& f_load;  delete& a;
delete& k;
// ======================================================================
// Determine Internal Forces N(ik) at Nodes G
// ======================================================================
//
std::cout << "Internal␣Forces" << endl;
std::cout << "===============" << endl << endl;
//
// Elasticity tensor E_cont(i,k,l,delta) =
//                               E_xy(i,k,l,m) * g_kont(m,delta)
// ------------------------------------------------------------------
MATRIX E_cont(dim_i, dim_i, dim_i, dim_n);
//
E_cont = E_xy.dimension(dim_i * dim_i * dim_i, dim_i) *
                              g_cont.dimension(1, dim_i, dim_n);
//
std::cout
    << "Transformed␣elasticity␣tensor␣E_cont(i,␣k,␣l,␣delta):"
    << E_cont << endl;
                                             outFILE << "" << endl;
// ---------------------------------------------------------------------
```

```
// Calculate internal forces n(i, K) at all nodes
// ------------------------------------------------
MATRIX n(dim_i, dim_i);
//
// n(i, k) = E_cont(i, k, l, delta) * v(l, delta)
//                             calculated with displacement vector
//
n = E_cont.dimension(dim_i * dim_i, dim_i * dim_P) *
                                    v.dimension(1, dim_i * dim_P);
//
std::cout
    << "Internal␣forces␣n(i,␣K)␣of␣all␣nodes:␣" << n << endl;
                                            outFILE << "" << endl;
    // ===============================================================
    std::cout << "End␣Triangle" << endl;
    std::cout << "============" << endl;
    //
    std::cout << "Pause:" << endl;
    // ###########################
    char c;
    std::cin >> c;                            outFILE << c << endl;
    // ===============================================================
    return 4;
};
// ===================================================================
```

As a numerical example for the triangular element, the lower left corner of the rectangle of Section 7.1 with dimensions of width 5 and height 10 with a base angle of 90 degrees is chosen as illustrated in Fig. 7.2. This is to demonstrate comparable results for the differential geometry. As a simplification for the transparent interpretation of the calculation results, again hypothetical values for the plate thickness $t = 1.0$, elasticity modulus $E = 1.0$, and Poisson's ratio $\nu = 0.0$ are assumed.

The calculation below was performed with a prescribed displacement field simulating pure tensile stress in the $\mathbf{e}^0$-direction to make the results easily verifiable. The specification of a uniform surface load is purely theoretical, aiming to illustrate the conversion into nodal loads of the load vector.
The results of the calculations are presented below:

Listing 7.5: Element Matrices of the TRIANGLE Element with Tension in x-Direction

```
Finite linear Triangle Element
==============================

Deformation in the x-direction due to Tension with given displacements
======================================================================

External node numbers P in internal order:
[ (1*3) * (1*1) ]-intMATRIX:
Columns 0 to 2
         0          1          2
Press -> Enter key:
```

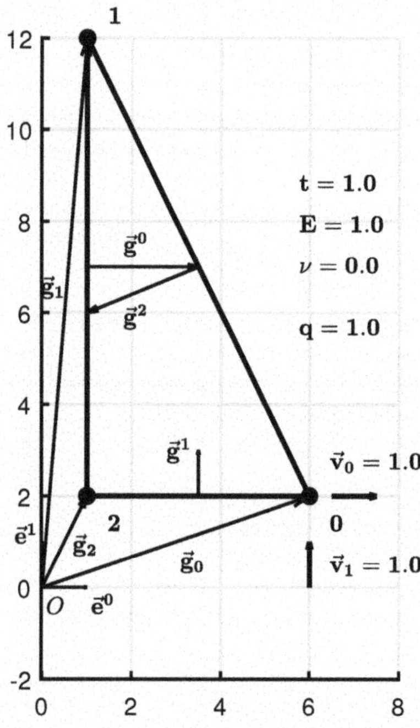

Fig. 7.2: Demonstration example of a linear triangular element under tension and shear, with an enlargement of the contravariant basis vectors $\mathbf{g}^{\alpha}$ by a factor of A/4

```
Natural coordinates theta(alpha,P) of nodes P:
[ (3*3) * (1*1)]-MATRIX:
Columns 0 to 2
 1.000e+00  0.000e+00  0.000e+00
 0.000e+00  1.000e+00  0.000e+00
 0.000e+00  0.000e+00  1.000e+00

Pause -> Enter a character:

Enter coordinates x(0)=x, x(1)=y of nodes P:
[ (2*3) * (1*1) ]-MATRIX:
Columns 0 to 2
 6.000e+00  1.000e+00  1.000e+00
 2.000e+00  1.200e+01  2.000e+00

Pause -> Enter a character:

Differential Geometry
=====================
```

```
Covariant basis vectors g:
[ (2*3) * (1*1) ]-MATRIX:
Columns 0 to 2
 6.000e+00  1.000e+00  1.000e+00
 2.000e+00  1.200e+01  2.000e+00

Pause -> Enter a character:

Covariant basis vectors a of the virtual tetrahedron:
[ (3*3) * (1*1) ]-MATRIX:
Columns 0 to 2
 6.000e+00  1.000e+00  1.000e+00
 2.000e+00  1.200e+01  2.000e+00
 1.000e+00  1.000e+00  1.000e+00

Pause -> Enter a character:

Covariant metric of a_cov of the virtual tetrahedron:
[ (3*3) * (1*1) ]-MATRIX:
Columns 0 to 2
 4.100e+01  3.100e+01  1.100e+01
 3.100e+01  1.460e+02  2.600e+01
 1.100e+01  2.600e+01  6.000e+00

Pause -> Enter a character:

e-Matrix:
[ (3*3) * (3*1) ]-MATRIX:
Columns 0 to 2
 0.000e+00  0.000e+00  0.000e+00
 0.000e+00  0.000e+00 -1.000e+00
 0.000e+00  1.000e+00  0.000e+00
 0.000e+00  0.000e+00  1.000e+00
 0.000e+00  0.000e+00  0.000e+00
-1.000e+00  0.000e+00  0.000e+00
 0.000e+00 -1.000e+00  0.000e+00
 1.000e+00  0.000e+00  0.000e+00
 0.000e+00  0.000e+00  0.000e+00

Pause -> Enter a character:

Area element = 50

Contravariant basis vectors a of the virtual tetrahedron:
[ (3*3) * (1*1) ]-MATRIX:
Columns 0 to 2
 2.000e-01  0.000e+00 -2.000e-01
 0.000e+00  1.000e-01 -1.000e-01
-2.000e-01 -2.000e-01  1.400e+00

Pause -> Enter a character:

Test: Orthogonality of basis vectors a of the virtual tetrahedron:
[ (3*3) * (1*1) ]-MATRIX:
```

```
Columns 0 to 2
 1.000e+00  0.000e+00  0.000e+00
 0.000e+00  1.000e+00  0.000e+00
 0.000e+00  0.000e+00  1.000e+00

Pause -> Enter a character:

Contravariant basis vectors g:
[ (2*3) * (1*1) ]-MATRIX:
Columns 0 to 2
 2.000e-01  0.000e+00 -2.000e-01
 0.000e+00  1.000e-01 -1.000e-01

Pause -> Enter a character:

Test: Sum of contravariant basis vectors g:
[ (2*1) * (1*1) ]-MATRIX:
Columns 0 to 0
 0.000e+00
 0.000e+00

Pause -> Enter a character:

Input: > 0 thickness /
       = 0  (default: only shear) /
       < 0  (default: nue = 0): -1

Plate thickness t: 1
Young's␣modulus␣E:␣1
Poisson's ratio nue: 0

Plane Strain State
==================

Elasticity matrix E(i,k,m,l):
[ (4*4) * (1*1) ]-MATRIX:
Columns 0 to 3
 1.000e+00  0.000e+00  0.000e+00  0.000e+00
 0.000e+00  5.000e-01  5.000e-01  0.000e+00
 0.000e+00  5.000e-01  5.000e-01  0.000e+00
 0.000e+00  0.000e+00  0.000e+00  1.000e+00

Pause -> Enter a character:

Transformed elasticity tensor E_contra(i,l,beta,delta):
[ (2*2) * (3*3) ]-MATRIX:
Columns 0 to 5
 4.000e-02  0.000e+00 -4.000e-02  0.000e+00  5.000e-03 -5.000e-03
 0.000e+00  0.000e+00  0.000e+00  1.000e-02  0.000e+00 -1.000e-02
 0.000e+00  1.000e-02 -1.000e-02  0.000e+00  0.000e+00  0.000e+00
 2.000e-02  0.000e+00 -2.000e-02  0.000e+00  1.000e-02 -1.000e-02

Columns 6 bis 8
-4.000e-02 -5.000e-03  4.500e-02
```

```
-1.000e-02  0.000e+00  1.000e-02
 0.000e+00 -1.000e-02  1.000e-02
-2.000e-02 -1.000e-02  3.000e-02

Pause -> Enter a character:

Symmetric elasticity tensor E_contra(i,beta,l,delta):
[ (2*3) * (2*3) ]-MATRIX:
Columns 0 to 5
 4.000e-02  0.000e+00  0.000e+00  0.000e+00 -4.000e-02  0.000e+00
 0.000e+00  2.000e-02  1.000e-02  0.000e+00 -1.000e-02 -2.000e-02
 0.000e+00  1.000e-02  5.000e-03  0.000e+00 -5.000e-03 -1.000e-02
 0.000e+00  0.000e+00  0.000e+00  1.000e-02  0.000e+00 -1.000e-02
-4.000e-02 -1.000e-02 -5.000e-03  0.000e+00  4.500e-02  1.000e-02
 0.000e+00 -2.000e-02 -1.000e-02 -1.000e-02  1.000e-02  3.000e-02

Pause -> Enter a character:

Determine Stiffness Matrix
==========================

Matrix M(K,M, beta,delta) without Factor 1/2:
[ (9*9) * (1*1) ]-MATRIX:
Columns 0 to 5
 1.000e+00  0.000e+00  0.000e+00  0.000e+00  0.000e+00  0.000e+00
 0.000e+00  1.000e+00  0.000e+00  0.000e+00  0.000e+00  0.000e+00
 0.000e+00  0.000e+00  1.000e+00  0.000e+00  0.000e+00  0.000e+00
 0.000e+00  0.000e+00  0.000e+00  1.000e+00  0.000e+00  0.000e+00
 0.000e+00  0.000e+00  0.000e+00  0.000e+00  1.000e+00  0.000e+00
 0.000e+00  0.000e+00  0.000e+00  0.000e+00  0.000e+00  1.000e+00
 0.000e+00  0.000e+00  0.000e+00  0.000e+00  0.000e+00  0.000e+00
 0.000e+00  0.000e+00  0.000e+00  0.000e+00  0.000e+00  0.000e+00
 0.000e+00  0.000e+00  0.000e+00  0.000e+00  0.000e+00  0.000e+00
Columns 6 to 8
 0.000e+00  0.000e+00  0.000e+00
 0.000e+00  0.000e+00  0.000e+00
 0.000e+00  0.000e+00  0.000e+00
 0.000e+00  0.000e+00  0.000e+00
 0.000e+00  0.000e+00  0.000e+00
 0.000e+00  0.000e+00  0.000e+00
 1.000e+00  0.000e+00  0.000e+00
 0.000e+00  1.000e+00  0.000e+00
 0.000e+00  0.000e+00  1.000e+00

Pause -> Enter a character:

Stiffness matrix k(i,K,l,M):
[ (2*3) * (2*3) ]-MATRIX:
Columns 0 to 5
 1.000e+00  0.000e+00  0.000e+00  0.000e+00 -1.000e+00  0.000e+00
 0.000e+00  5.000e-01  2.500e-01  0.000e+00 -2.500e-01 -5.000e-01
 0.000e+00  2.500e-01  1.250e-01  0.000e+00 -1.250e-01 -2.500e-01
 0.000e+00  0.000e+00  0.000e+00  2.500e-01  0.000e+00 -2.500e-01
-1.000e+00 -2.500e-01 -1.250e-01  0.000e+00  1.125e+00  2.500e-01
```

```
 0.000e+00 -5.000e-01 -2.500e-01 -2.500e-01  2.500e-01  7.500e-01

Pause -> Enter a character:

Distributed Loads
=================

Enter surface loads q(0)=q(x), q(1)=q(y) of the nodes L enter
-> 1 = yes / 0 = (zero) / < 0 (uniform load q(i)):
-1
Uniform load q(i) for all nodes:

Enter [ (1*2) * (1*1) ]-MATRIX (row-wise !!!):
Input done!
1 1

Surface loads q(i,L) of the nodes L:[
(2*3) * (1*1) ]-MATRIX:
Columns 0 to 2
 1.000e+00  1.000e+00  1.000e+00
 1.000e+00  1.000e+00  1.000e+00

Pause -> Enter a character:

Load vector f_load(i,K):
[ (2*3) * (1*1) ]-MATRIX:
Columns 0 to 2
 8.333e+00  8.333e+00  8.333e+00
 8.333e+00  8.333e+00  8.333e+00

Pause -> Enter a character:

Displacement State
==================

Enter displacements v(0)=v(x), v(1)=v(y) of nodes
-> 1 = yes / 0 = (null) / < 0 (Test):
 /-1 (x-Tension) /-2 (y-Tension) /-3 (x-Shear) /-4 (y-Shear):
-1

Displacement vector v(l, M) of the nodes M:
[ (2*3) * (1*1) ]-MATRIX:
Spalten 0 bis 2
 1.000e+00  0.000e+00  0.000e+00
 0.000e+00  0.000e+00  0.000e+00

Pause -> Enter a character:

Force vector f(i, K) = k(i, k, l, m) * v(l, m):
[ (2*3) * (1*1) ]-MATRIX:
Spalten 0 bis 2
 1.000e+00  0.000e+00 -1.000e+00
 0.000e+00  0.000e+00  0.000e+00
```

```
Pause -> Enter a character:

Reaction vector a(i, K) = k(i, k, l, m) * v(l, m) - f_load(i, K):
[ (2*3) * (1*1) ]-MATRIX:
Spalten 0 bis 2
-7.333e+00 -8.333e+00 -9.333e+00
-8.333e+00 -8.333e+00 -8.333e+00

Pause -> Enter a character:

Internal Forces
===============

Transformed elasticity tensor E_cont(i,k,l,delta):
[ (2*2) * (2*3) ]-MATRIX:
Columns 0 to 5
 2.000e-01  0.000e+00  0.000e+00  0.000e+00 -2.000e-01  0.000e+00
 0.000e+00  1.000e-01  5.000e-02  0.000e+00 -5.000e-02 -1.000e-01
 0.000e+00  1.000e-01  5.000e-02  0.000e+00 -5.000e-02 -1.000e-01
 0.000e+00  0.000e+00  0.000e+00  1.000e-01  0.000e+00 -1.000e-01

Pause -> Enter a character:

Internal forces n(i,K) of all nodes:
[ (2*2) * (1*1) ]-MATRIX:
Columns 0 to 1
2.000e-01  0.000e+00
0.000e+00  0.000e+00

Pause -> Enter a character:

End Triangle
============
```

To verify the results, a second calculation for the same system is performed with a modified displacement field to generate pure shear deformation in the $\mathbf{e}^1$-direction. The abbreviated listing below displays the results:

Listing 7.6: Element Matrices of the TRIANGLE Element with Shear in y-Direction

```
Finite linear Triangle Element
===============================

Deformation in the y-direction due to Shear with given displacements
====================================================================

External node numbers P in internal order:
[ (1*3) * (1*1) ]-intMATRIX:
Spalten 0 bis 2
        0          1          2

Pause -> Enter a character:

.......
```

```
Determine Stiffness Matrix
==========================

Matrix M(K,M, beta,delta) without factor 1/2:
[ (9*9) * (1*1) ]-MATRIX:
Columns 0 to 5
 1.000e+00  0.000e+00  0.000e+00  0.000e+00 -1.000e+00  0.000e+00
 0.000e+00  5.000e-01  2.500e-01  0.000e+00 -2.500e-01 -5.000e-01
 0.000e+00  2.500e-01  1.250e-01  0.000e+00 -1.250e-01 -2.500e-01
 0.000e+00  0.000e+00  0.000e+00  2.500e-01  0.000e+00 -2.500e-01
-1.000e+00 -2.500e-01 -1.250e-01  0.000e+00  1.125e+00  2.500e-01
 0.000e+00 -5.000e-01 -2.500e-01 -2.500e-01  2.500e-01  7.500e-01

Pause -> Enter a character:

Displacement State
==================

Enter displacements v(0)=v(x), v(1)=v(y) of nodes
-> 1 = yes / 0 = (zero) / < 0 (Test):
   / -1 (x-Tension) / -2 (y-Tension) /-3 (x-Shear) / -4 (y-Shear):
-4
Displacement vector: Shear in y-direction

Displacement vector v(l,M) of the nodes M:
[ (2*3) * (1*1) ]-MATRIX:
Columns 0 to 2
 0.000e+00  0.000e+00  0.000e+00
 1.000e+00  0.000e+00  0.000e+00

Pause -> Enter a character:

Force vector f(i,K) = k(i,k,l,m) * v(l,m):
[ (2*3) * (1*1) ]-MATRIX:
Columns 0 to 2
 0.000e+00  2.500e-01 -2.500e-01
 5.000e-01  0.000e+00 -5.000e-01

Pause -> Enter a character:

Internal Forces
===============

Internal forces n(i,K) of all nodes:
[ (2*2) * (1*1) ]-MATRIX:
Columns 0 to 1
 0.000e-01  1.000e+00
 1.000e+00  0.000e+00

Pause -> Enter a character:

End Triangle
============
```

The properties of the triangular element can be verified and examined based on the numerical results of both calculations, according to the elementary equilibrium and deformation checks of mechanics/statics, along with the unit displacement states contained in the stiffness matrix.

Internet Links

Meissner 2023. Meißner, U. F.: Object-oriented Tensor and Matrix Classes (2023)
https://rebrand.ly/Matrix_Classes

Chapter 8
Conclusion

With the present concept, a comprehensive synthesis from theoretical tensor notation and traditional matrix calculus to new classes of matrices and object-oriented algorithms for mechanics and engineering problems was accomplished. This integral approach to the elementary tensor calculus of tensor bases has been realized with the tensor class BASIS, and was tested using the function METRIC. In this context, also graphical methods for the perspective representation of basis vectors and tensor quantities play a significant role in illustrating the correlation of these quantities in three-dimensional space. For this purpose, the graphic classes PERSPEKTIVE and PyPLOT were created and utilized with the function BASES_3D for three-dimensional stereo vision.

The classical tensor calculus is well known and provides a proven theoretical basis for the formulation of many mathematical-physical relationships in science and engineering. Within the finite element method, it is needed in many respects for the differential geometry and stiffness relationships of plane and spatial structures. In particular, it is essential for spatially curved structures, as different reference systems have to be managed, and the consistency of the transformations of the mechanical quantities is of main importance. Conclusively, the advantage of the presented approach was illustrated by exemplary applications of the finite element method, namely the functions PARALLELOGRAM and TRIANGLE for the generation of typical elements.

For the reliable realization of such tasks and for the efficient numerical computation of the discretized FEM systems, the conventional matrix calculus is known and variously proven. Traditionally, it is applied by use of symbolic notation and two-dimensional calculation schemes, as described in numerous publications and program libraries. Less known and common is the handling of multidimensional matrices in index notation. This is probably due to the fact that the variety and number of indices are considered by many users a priori to be relatively confusing, as in the case of the corresponding tensor notation. However in teaching, it turned out that this is mainly a problem of didactic and a question of accustoming to this kind of notation. Experience shows that young students are rather quick at learning the index notation, and also gain a good overview and master its handling after a short time. The key to success lies primarily in consistent practice,

U. F. Meissner, *Tensor Calculus with Object-Oriented Matrices for Numerical Methods in Mechanics and Engineering*, https://doi.org/10.1007/978-3-031-59302-4_8

especially under appropriate guidance. This book was intended to get familiar with the notation, even by the initial bold print of the summation notations.

However, it is not widely known that the index notation can also be used to transform tensor theory directly into operational numerical methods with multidimensional matrices. For these purposes a corresponding mapping of the matrix calculus to multidimensional arrays of the programming language is necessary. But due to a lack of adequate syntax and semantics for matrix objects in the past, a deficit exists up to now in the mathematical-scientifically oriented programming languages. This may be overcome by the use of object-oriented methods. The book shows that this approach has been used successfully to make a seamless transition from the well-established matrix calculus to numeric with object-oriented matrices. The context is deliberately limited to four-dimensional rectangular matrices in order not to jeopardize the conciseness of the presentation by a too large variety of special matrix forms. The extension of the presented matrix classes is of course possible in every respect.

With the program and calculation examples described, the essential characteristics of the object-oriented methodology for the realization of the synthesis of matrices, tensors, software objects and numerical algorithms were represented and tested. From this approach, great advantages arise in practical applications in terms of clarity and reliability of the corresponding program structures and of the numeric algorithms, which can be summarized as follows. The encapsulation principle for the internal matrix structures and methods provides high security against the destruction of data properties of the generated matrix objects and against the misuse of private methods, both during programming and run time. The demonstration examples highlight the advantages of the inheritance strategy in class properties and methods, for example, from the abstract base class MATRIZEN to the derived classes MATRIX, VEKTOR, e_MATRIX, and delta_MATRIX. The binding of methods (member functions) to the generated class instances (objects) prevents misuse and unauthorized access by foreign objects. *Overloading the standard operators of C++ provides the innovative possibility to define comparable matrix operations and use them in a compliant manner for processing of matrix objects.* With the supplemented matrix class qMATRIX, even the resolution of linear equation systems could be optionally formulated in a compact manner.

With the material presented, the author aims to encourage readers to pursue further study in object-oriented methods in general and numerical techniques with matrix objects in particular. As the object-oriented paradigm has been established since the 1980s in the analysis and modeling of processes [Coad/Yourdon 1990], in software design and programming [Breymann 1997], and as a fundamental modeling and design method [Rumbaugh/Blaha/Premerlani/Eddy/Lorensen 1991], it has proven itself in various applications and is now internationally recognized as a standard, such as in **B**uilding **I**nformation **M**odeling (BIM) [Borrmann/König/Koch/Beetz (Hrsg.) 2018].

References

Borrmann/König/Koch/Beetz (Hrsg.) 2018. Borrmann A.; König M.; Koch Chr.; Beetz J.: Building Information Modeling: Technology Foundations and Industry Practice. Springer, Cham (2018)

Breymann 1997. Breymann, U.: C++ - Eine Einführung. 4. Aufl., Carl Hanser, München (1997)

Coad/Yourdon 1990. Coad, P.; Yourdon, E.: Object-Oriented Analysis. Prentice Hall, Englewood Cliffs (1990)

Rumbaugh/Blaha/Premerlani/Eddy/Lorensen 1991. Rumbaugh, J.; Blaha, M.; Premerlani, W.; Eddy, F.; Lorensen, W.: Object-Oriented Modeling and Design. Prentice Hall, Englewood Cliffs (1991)

Chapter 9
Listings

References

Baker 1991. Baker, A. J.; Pepper, D. W.: Finite Elements 1-2-3. McGraw-Hill, New York (1991)

Bathe 1986. Bathe, K.-J.: Finite-Elemente-Methoden. Springer, Berlin (1986)

Beyer 1977. Beyer, A.: Berechnung großflächiger Grundwasserströmungen mit Vertikalstruktur mit Hilfe der Finte Element Methode. Fortschritt-Berichte der VDI Zeitschriften Nr. 34, Reihe 4 Bauingenieurwesen, VDI-Verlag, Düsseldorf (1977)

Borrmann/König/Koch/Beetz (Hrsg.) 2018. Borrmann A.; König M.; Koch Chr.; Beetz J.: Building Information Modeling: Technology Foundations and Industry Practice. Springer, Cham (2018)

Breymann 1993. Breymann, U.: C++ - Eine Einführung. Carl Hanser, München (1993)

Breymann 1997. Breymann, U.: C++ - Eine Einführung. 4. Aufl., Carl Hanser, München (1997)

Clough/Penzien 1975. Clough, R. W.; Penzien, J.: Dynamics of Structures. McGraw-Hill, New York (1975)

Coad/Yourdon 1990. Coad, P.; Yourdon, E.: Object-Oriented Analysis. Prentice Hall, Englewood Cliffs (1990)

de Boer/Schroeder 2012. de Boer, R.; Schröder, J.: Tensor Calculus for Engineers: Analytical and Computational Aspects. Springer, Berlin (2012)

Duschek/Hochrainer 1968. Duschek, A.; Hochrainer, A.: Grundzüge der Tensorrechnung in analytischer Darstellung. Springer, Wien (1968)

Ebeling 1977. Ebeling, H.: Berechnung der Vertikalstruktur wind- und gezeitenerzeugter Strömungen nach der Methode der finiten Elemente. Fortschritt-Berichte der VDI Zeitschriften Nr. 32, Reihe 4 Bauingenieurwesen, VDI-Verlag, Düsseldorf (1977)

Ehlers 1971. Ehlers, K.-D.: Berechnung instationärer Grund- und Sickerwasserströmungen mit freier Oberfläche nach der Methode finiter Elemente. Dissertation, Institut für Strömungsmechanik, Universität Hannover (1971)

Green/Zerna 1954. Green, A. E.; Zerna, W.: Theoretical Elasticity. At The Clarendon Press, Oxford (1954)

Grotkop 1972. Grotkop, G.: Die Berechnung von Flachwasserwellen nach der Methode der finiten Elemente. Dissertation, Institut für Strömungsmechanik, Universität Hannover (1972)

Hackbusch 1991. Hackbusch, W.: Iterative Lösung großer schwachbesetzter Gleichungssysteme. Teubner, Stuttgart - Studienbücher Mathematik (1991)

Hartmann (Hrsg.) 2000. Hartmann, D.: Objektorientierte Modellierung in Planung und Konstruktion - DFG-Bericht, Wiley-VCH, Weinheim (2000)

Herrling 1977. Herrling, B.: Eine hybride Formulierung in Wasserständen zur Berechnung von Flachwasserwellen mit der Methode finiter Elemente. Fortschritt-Berichte der VDI Zeitschriften Nr. 37, Reihe 4 Bauingenieurwesen, VDI-Verlag, Düsseldorf (1977)

U. F. Meissner, *Tensor Calculus with Object-Oriented Matrices for Numerical Methods in Mechanics and Engineering*, https://doi.org/10.1007/978-3-031-59302-4_9

Herrling/Pfeiffer/Beyer 1979. Herrling, B.; Pfeiffer, E., Beyer A.: Das Programmsystem MECCA, ein modulares Elementkonzept für die Kontinuumsanalysis. Angewandte Informatik, S. 17-22 (1/1979)
Klingbeil 1966. Klingbeil, E.: Tensorrechnung für Ingenieure. Bibliographisches Institut, Mannheim (1966)
Laemmer 1996. Lämmer, L.: Parallelisierung von Anwendungen der Finite-Element-Methode im Bauingenieurwesen. Habilitationsschrift TU Darmstadt (1996)
Laemmer/Meissner/Ruben 2003. Lämmer L., Meißner U., Ruben J.: Finite Element Modelling of Soil-Structure Systems in Workstation Clusters. Proceedings Second MIT Conference on Computational Fluid and Solid Mechanics, Vol. 2, P. 2312, Elsevier Cambridge (2003)
Meissner 1971. Meißner, U.: Berechnung von Schalen unter großen Verschiebungen und Verdrehungen bei kleinen Verzerrungen mit Hilfe finiter Dreieckselemente. Dissertation, Institut für Strömungsmechanik, Universität Hannover (1971)
Meissner 1978. Meißner, U.: MEDYNA I - Ein maschinenunabhängiges FORTRAN-Programmsystem zur statischen Berechnung von Stab- und Flächentragwerken mit der Methode der finiten Elemente. Fortschritt-Berichte der VDI Zeitschriften Nr. 41, Reihe 4 Bauingenieurwesen, VDI-Verlag, Düsseldorf (1978)
Meissner 1996. Meißner, U.: Ein übergeordnetes Variationsprinzip für den Fehlerabgleich der Finite-Element-Methode. Technische Mechanik 16, S. 333 (1996)
Meissner 2022. Meißner, U. F.: Tensorkalkül mit objektorientierten Matrizen für numerische Methoden in Mechanik und Ingenieurwissenschaften - Eine Grundlage für Tensor-/Matrix-Algorithmen der Finite-Elemente-Methode. 1st Edition - Springer Vieweg, Wiesbaden (2022)
Meissner/Maurial 2000. Meißner, U. F.; Maurial, A.: Die Methode der Finiten Elemente - Eine Einführung in die Grundlagen. Springer, Berlin (2000)
Pestel/Leckie 1963. Pestel, E. C.; Leckie, F. A.: Matrix Methods in Elastomechanics. McGraw-Hill, New York (1963)
Rueppel (Hrsg.) 2007. Rüppel, U.: Vernetzt-kooperative Planungsprozesse im konstruktiven Ingenieurbau : Grundlagen, Methoden, Anwendung und Perspektiven zur vernetzten Ingenieurkooperation. Springer, Berlin (2007)
Rumbaugh/Blaha/Premerlani/Eddy/Lorensen 1991. Rumbaugh, J.; Blaha, M.; Premerlani, W.; Eddy, F.; Lorensen, W.: Object-Oriented Modeling and Design. Prentice Hall, Englewood Cliffs (1991)
Schulze 1973. Schulze, K.-W.: Eine problemorientierte Sprache für die Dynamik offener Gerinne (DOG). Dissertation, Institut für Strömungsmechanik, Universität Hannover, Mitteilungen SFB 79, Heft 1 (1973)
Withum / SFB 79 1973. Rechnergruppe: TAS 86 - Programmdokumentation, Eigenverlag SFB79, TU Hannover (1973)
Zienkiewicz/Taylor 1989. Zienkiewicz, O. C.; Taylor, R. L.: The Finite Element Method. McGraw-Hill, London (1989)
Zurmuehl/Falk 1986. Zurmühl, R.; Falk, S.: Matrizen und ihre Anwendungen. Springer, Berlin (1986)

Internet Links

IIB. Institut für Numerische Methoden und Informatik im Bauwesen, TU Darmstadt (2021)
`https://www.iib.tu-darmstadt.de/institut_iib/index.de.jsp`
Kacerovsky 2020. Kacerovsky, J. Benjamin: DrawVector (2020)
`https://github.com/JBKacerovsky/Draw_Vector`
MATLAB R2022a. MathWorks: Produktfamilie (2022)
`https://de.mathworks.com/products/matlab.html`
Meissner 2006. Meißner, U. F.: Schriften (2006)
`https://ing-buero.meissner.info/Schriften`
Meissner 2023. Meißner, U. F.: Object-oriented Tensor and Matrix Classes (2023)
`https://rebrand.ly/Matrix_Classes`

Microsoft 2023. Visual Studio - Visualizing data with mathplotlib (2023)
https://learn.microsoft.com/en-us/search/?terms=matplotlib-cpp-visual-studio
https://stackoverflow.com/questions/66507618/
running-python-matplotlibcpp-in-visual-studio-2019
Zuse 2019. Zuse, H.: Konrad Zuse Multimedia Show (2018/19)
http:
//www.horst-zuse.homepage.t-online.de/_wp_generated/zuse-broschuere.html

Figures [1, 2]

[1] The depicted figures were created using [MATLAB R2022a], the software platform for graphics, numerical computations, and programming.
[2] In this process, the DrawVector function - available for MATLAB on the WWW - was utilized as presented by [Kacerovsky 2020].

Tables

Programs

Subject Index

U. F. Meissner, *Tensor Calculus with Object-Oriented Matrices for Numerical Methods in Mechanics and Engineering*, https://doi.org/10.1007/978-3-031-59302-4

The manufacturer's authorised representative in the EU is Springer Nature Customer Service Centre GmbH, Europaplatz 3, 69115 Heidelberg, Germany. If you have any concerns regarding our products, please contact ProductSafety@springernature.com

Printed and bound by CPI Group (UK) Ltd, Croydon, CR0 4YY
15/07/2026
02167632-0007